BK
SY

22183

£29-00

NUTRITIVE SWEETENERS

An industry–university co-operation Symposium organised under the auspices of the National College of Food Technology, University of Reading, 30 March–1 April, 1981

THE SYMPOSIUM COMMITTEE

Mrs A. ALFORD,
Manager, Library and Information Services, Tate & Lyle Ltd, Group Research & Development, Philip Lyle Memorial Research Laboratory, PO Box 68, Reading, Berks. RG6 2BX.

GORDON G. BIRCH, B.Sc., Ph.D., D.Sc. (Lond), F.R.S.C., C. Chem., M.R.S.H.
Reader at National College of Food Technology, Reading University, Weybridge, Surrey KT13 0DE.

C. BUCKE, B.Sc., Ph.D.
Programme Manager, Biotechnology and Food Products Research, Tate & Lyle Ltd, Group Research & Development, Philip Lyle Memorial Research Laboratory, PO Box 68, Reading, Berks. RG6 2BX.

J. A. FORD, B.A. (Oxon)
Secretary at National College of Food Technology, Reading University, Weybridge, Surrey KT13 0DE.

K. J. PARKER, M.A., D.Phil. (Oxon), F.R.S.C., C. Chem.
Chief Scientist, Tate & Lyle Ltd, Group Research & Development, Philip Lyle Memorial Research Laboratory, PO Box 68, Reading, Berks. RG6 2BX.

E. J. ROLFE, B.Sc., M. Chem. A., F.R.S.C., C. Chem., F.I.F.S.T.
Principal, National College of Food Technology, Reading University, Weybridge, Surrey KT13 0DE.

Mrs B. A. SHORE,
National College of Food Technology, Reading University, Weybridge, Surrey KT13 0DE.

NUTRITIVE SWEETENERS

Edited by
G. G. BIRCH and K. J. PARKER

APPLIED SCIENCE PUBLISHERS
LONDON and NEW JERSEY

APPLIED SCIENCE PUBLISHERS LTD
Ripple Road, Barking, Essex, England

APPLIED SCIENCE PUBLISHERS INC.
Englewood, New Jersey 07631, USA

British Library Cataloguing in Publication Data

Nutritive sweeteners.
1. Food additives—Congresses
2. Flavoring essences—Congresses
3. Natural sweeteners—Congresses
I. Birch, G. G.   II. Parker, K. J.
641.3′38   TP418

ISBN 0-85334-997-5

WITH 49 TABLES AND 68 ILLUSTRATIONS

© APPLIED SCIENCE PUBLISHERS LTD 1982

The selection and presentation of material and the opinions expressed in this publication are the sole responsibility of the authors concerned.

All rights reserved. No part of this publication may be reproduced, stored in a retrieval system, or transmitted in any form or by any means, electronic, mechanical, photocopying, recording, or otherwise, without the prior written permission of the publishers, Applied Science Publishers Ltd, Ripple Road, Barking, Essex, England

Photoset in Malta by Interprint Limited
Printed in Great Britain by Galliard (Printers) Ltd, Great Yarmouth

List of Contributors

G. G. BIRCH

National College of Food Technology, University of Reading, St. George's Avenue, Weybridge, Surrey KT13 0DE, UK.

W. M. EDGAR

University of Newcastle upon Tyne, Dental School, Framlington Place, Newcastle upon Tyne NE2 4BW, UK.

A. FAURION

Laboratoire de Neurobiologie Sensorielle, CENFAR BP6, 92269 Fontenay aux Roses, France.

P. D. FULLBROOK

National College of Food Technology, University of Reading, St. George's Avenue, Weybridge, Surrey KT13 0DE, UK.

L. HYVÖNEN

Department of Food Chemistry and Technology, University of Helsinki, SF-00710 Helsinki 71, Finland.

M. R. KARE

Monell Chemical Senses Center, 3500 Market Street, Philadelphia, Pennsylvania 19194, USA

M. W. KEARSLEY

National College of Food Technology, University of Reading, St. George's Avenue, Weybridge, Surrey KT13 0DE, UK.

P. KOIVISTOINEN

Department of Food Chemistry and Technology, University of Helsinki, SF-00710 Helsinki 71, Finland.

R. H. P. LIAN-LOH

National College of Food Technology, University of Reading, St. George's Avenue, Weybridge, Surrey KT13 0DE, UK.

P. LINKO

Department of Chemistry, Helsinki University of Technology, 02150 Espoo 15, Finland.

I. MACDONALD

Department of Physiology, Guy's Hospital Medical School, London SE1 9RT, UK.

P. MACLEOD

Laboratoire de Neurobiologie Sensorielle, CENFAR BP6, 92269 Fontenay aux Roses, France.

S. L. MUNTON

National College of Food Technology, University of Reading, St. George's Avenue, Weybridge, Surrey KT13 0DE, UK.

M. NAIM

The Hebrew University of Jerusalem, The Levi Eshkol School of Agriculture, Faculty of Agriculture, Rehovot 76-100, PO Box 12, Israel.

W. M. NICOL

Tate & Lyle Limited, Group Research and Development, Philip Lyle Memorial Research Laboratory, PO Box 68, Reading, Berks. RG6 2BX, UK.

G. OGUNMOYELA

National College of Food Technology, University of Reading, St. George's Avenue, Weybridge, Surrey KT13 0DE, UK.

T. J. PALMER

Tunnel Refineries Limited, Thames Bank House, Tunnel Avenue, Greenwich, London SE10 0PA, UK.

E. J. ROLFE

National College of Food Technology, University of Reading, St. George's Avenue, Weybridge, Surrey KT13 0DE, UK.

NANCY J. ROTHWELL

Department of Physiology, St. George's Hospital Medical School, Cranmer Terrace, Tooting, London SW17 0RE, UK.

SUSAN S. SCHIFFMAN

Department of Psychiatry, Duke Medical Center, Durham, North Carolina 17710, USA.

R. S. SHALLENBERGER

Department of Food Science and Technology, Cornell University, New York State Agricultural Experimental Station, Geneva, New York 14456, USA.

P. J. SICARD

Roquette Frères, 62136 Lestrem, France.

M. J. STOCK

Department of Physiology, St. George's Hospital Medical School, Cranmer Terrace, Tooting, London SW17 0RE, UK.

P. H. WIGGALL

International Scientific Standards, Cadbury Schweppes Limited, Bournville, Birmingham B30 2LU, UK.

Contents

	Page
List of Contributors	v
Introduction E. J. Rolfe	1
Paper 1. Keynote: Intrinsic Chemistry of the Nutritive Sweeteners R. S. Shallenberger	7
Paper 2. Sucrose, the Optimum Sweetener W. M. Nicol	17
Paper 3. The Use of Sugars in Confectionery P. H. Wiggall	37
Paper 4. Malt and Maltose Syrups P. D. Fullbrook	49
Paper 5. Nutritive Sweeteners from Starch T. J. Palmer	83
Paper 6. Lactose and Lactitol P. Linko	109
Paper 7. Fructose in Food Systems L. Hyvönen and P. Koivistoinen	133

Paper 8. *Hydrogenated Glucose Syrups, Sorbitol, Mannitol and Xylitol.* 145
P. J. Sicard

Paper 9. *Nutritional Significance of Sweetness* . . . 171
M. Naim and M. R. Kare

Paper 10. *Chemical and Technological Modification of Physiological Effects in Food Carbohydrates* . . . 195
M. W. Kearsley and R. H. P. Lian-Loh

Paper 11. *Sugars and Dental Caries* 205
W. M. Edgar

Paper 12. *The Body Weight Response to Nutritional Sweeteners* 225
I. Macdonald

Paper 13. *Obesity, Thermogenesis and Carbohydrate Metabolism.* 231
M. J. Stock and Nancy J. Rothwell

Paper 14. *Sweet Taste Receptor Mechanisms* . . . 247
A. Faurion and P. MacLeod

Paper 15. *Synergism and the Sweet Response* . . . 275
G. G. Birch, G. Ogunmoyela and S. L. Munton

Paper 16. *Multidimensional Concepts in Sweetness Evaluation* 287
Susan S. Schiffman

Index 311

Introduction

E. J. ROLFE
*National College of Food Technology,
University of Reading, Weybridge,
Surrey, UK*

It is only during the last 100 years or so that there has occurred a rapid increase in sugar (sucrose) consumption in industrialised countries resulting in the development of a large international trade in sugar, with the economy of some developing countries being almost entirely dependent on the sugar cane as a cash crop. The trade build-up in sugar is not dependent on man's hunger but on his sweet tooth and the desire he has to satisfy this need. Sucrose is not the ideal sweetener for all purposes and as a consequence, alternatives are now available to the food industry, particularly glucose syrups when one is taking special note of nutritive sweeteners.

The title of the symposium may be separated into its component parts, nutrition and sweeteners, and the papers to be presented may be conveniently divided up in this way. Both topics represent areas in which our knowledge unfortunately is all too scant. Let us first consider the aspect of nutrition.

When considering nutritive sweeteners we are essentially looking at carbohydrates which serve as an energy source in the diet. But man is not like a mechanical engine which obtains its energy by a relatively simple combustion process. In man the digestion of food and release of energy takes place in the alimentary canal, a musculo-membranous tube about 30 ft long extending from mouth to anus. It is lined throughout its length by mucous membranes and has poured into it at various stages a complex of enzymes and of hormones to regulate and perform the digestive processes. Hence the effectiveness with which the body can utilise the ingested food will depend on the adequacy or otherwise of the supply of enzymes and gastro-intestinal hormones, and deficiencies are not uncommon, e.g. diabetes mellitus.

The absorption of food is further complicated by the fact that man is a microcosm; his digestive tract is populated by a variety of microorganisms. Some are symbiotic, but others can be detrimental. In addition the microflora of the gut will be affected *inter alia* by diet and by drugs taken for therapeutic purposes.

Immediately the food is placed in the mouth the digestion process starts through comminution and admixing with enzymes. Also at this stage, particularly in the presence of sucrose, dental caries may be initiated, mineral matter from the teeth being dissolved by acid formed from the sugar by bacteria residing in the sticky dental plaque. Dental caries has a high incidence among the population and is one reason why there is pressure to reduce sucrose consumption.

However, our present knowledge of nutrition is limited to the 'nuts and bolts' of the subject. We can define the approximate requirements for, for example, calories and how these requirements are influenced by age, activity, etc, and we can explain the necessity for vitamins in the diet if good health is to be maintained in terms of their role in intermediary metabolism. Overt under-nutrition may be rare in the UK, but the possibility cannot be ignored that sub-optimal nutrition in respect of one or more nutrients may not be uncommon. The effects may be of considerable importance overall in relation to general health and well-being. All too little is known of what constitutes optimal nutrition and one is not helped in reaching a definition by measuring parameters such as weight or height, if one wishes to discover say the optimum growth rate for the human child. What is definite is that the age of menarche in girls is progressively becoming less, and during this century, for one reason or another, diet being at least a contributory factor, girls are developing much more rapidly. Do we know if this is a desirable objective?

Starvation and overfeeding can be detected readily, but somewhere within such boundaries lies the optimum. But what is it in precise terms? Little is known of the intake of energy, protein, minerals, vitamins, etc., which is optimal for health and development at different ages and in different physiological states. Many people will be familiar with the controversy regarding the amount of vitamin C we should ingest daily.

We are still only feeling our way forward to elucidating why some individuals manage on relatively few calories—Bender has observed Nepalese porters apparently working on only 1000 calories per day—whereas some people are able seemingly to eat in excess of what may be considered normal requirements and do not become obese.

Research is required to study the variation in energy needs between individuals on the mechanism of thermogenesis and its role in regulation, and on the physiology of the hypothalamus in relation to the control of appetite.

It has been said that we are the food that we eat and increasingly the truth of this statement is becoming apparent. Changes in dietary patterns, associated particularly with increasing affluence, are thought to contribute to the observed increase in the incidence of many diseases that now pose major health challenges. For example, it is now believed that the modern civilised diet contains an excess of sugar and fat, and that we should therefore increase our consumption of starch at the expense of sugar and some fat; however, we can only conjecture as to what is the desirable ratio of carbohydrate to fat in the food we ingest.

The role of nutritional factors in relation to, for example, cardiovascular disease, diabetes mellitus, various cancers, and diseases of the large bowel, requires careful investigation. We have been advised to eat less saturated fats and to include more fibre in our diet, and have been informed of the harm caused to the nutritive value of unsaturated fats due to *cis*, *trans* isomerism arising during partial hydrogenation. Evidently changes in the nature of our food intake must be viewed with caution, and unfortunately it is difficult to decide whether damage is occurring at an early stage, as any adverse effects may not become apparent except after ingesting the food for a considerable period of years. Is there any cause for concern in the use of hydrogenated sugars if they are to be consumed in substantial quantities by individuals?

The general public are becoming more concerned about the food which they eat on perhaps two health grounds. One is their wish to select and control their food intake to avoid obesity—the major instance of malnutrition in the western countries. Obesity is associated with diminished performance and increased morbidity and mortality from various diseases. The other concern is what they consider to be undue interference with our food, for example, the use of food additives, and the treatments given to food usually to make it appear more attractive, as in the production of white rather than wholemeal flour and the polishing of rice—the classic cause of beri-beri in developing countries. Apparently sucrose suffers from disadvantages on both these counts. It is a highly purified food, providing what some nutritionists term 'empty calories', and its consumption is linked in the minds of many people with obesity and tooth decay. But without doubt sucrose is a useful nutrient and is an essential ingredient of many of our most attractive and palatable foods. Indeed it is the success of the food industry in this direction, by making readily available cheap and palatable sweet foods, that persuades people to eat an excess and contributes to the obesity problem.

Turning now to sweetness, this is a fundamental gustatory response which still can be assessed and measured only by organoleptic tests. There is no instrument that can measure sweetness as, for example, the pH meter can measure acidity. The fundamental study of sweetness is being pursued in two directions:

1. A physiological approach to provide an understanding of the detection of the sensation of sweetness through the interaction of the sweet substance with taste cells and of the action of the nerve fibres that innervate them.
2. A chemical approach which attempts to relate structure and molecular configuration with sweetness and its intensity.

Both approaches are dealt with in symposium papers and although the subjects are difficult, considerable progress is being made; it must be said, however, that our knowledge is still rudimentary. It is thought that about 50 taste cells are grouped together to form a taste-bud. Is it possible that taste cells are preferentially selective to the sugars which they detect and might this explain at least in part the observed difference in the intensity of the sweet taste of sugars? Again the quality of sweetness varies (equality not to be confused with intensity), as, for example, between fructose and maltose. Could this also be explained in part on similar grounds?

Substances of widely different structure and constitution elicit a sweet response: for example, L-aspartyl-L-phenylalanine methyl ester is 100–200 times as sweet as sucrose. Theories regarding sweet taste must be sufficiently comprehensive to include the whole range of known sweeteners.

Again some compounds, such as sucrose, give a rapid sweet taste impact followed by a sharp cut off, whereas others give a lingering taste that may persist for some appreciable time. A theory has been put forward to explain this behaviour but it is far from being a proven fact. As we are becoming more aware of the properties and characteristics of sweeteners we are able to choose the one which most completely satisfies the specification or performance required.

Ultimately perhaps we shall be able to design and synthesise the ideal sweet-tasting compound for use in a particular situation, though that ideal is a long way off. Even were it possible, we may be unable to make use of it because of the very large cost of establishing safety in use through toxicity tests. That leads us on to another problem: is there no cheaper and quicker method of determining toxicity, such as the use of tissue culture? But that is a different story.

I hope I have adequately set the scene against which the papers of this symposium may now be viewed.

1

Keynote: Intrinsic Chemistry of the Nutritive Sweeteners

R. S. SHALLENBERGER
*Department of Food Science and Technology, Cornell University,
New York State Agricultural Experimental Station
Geneva, New York, USA*

ABSTRACT

The intrinsic chemistry of the nutritive sweeteners (the sugars) is presented as a continuing problem in relating chiroptical properties to biological activity, in this case, sweetness. An interrelation between chiroptical properties, structure specification, and sweetness is herein developed.

INTRODUCTION

Mannose was one of the key reference compounds whose properties and reactions assisted in deducing the configurational structure of D-glucose and D-fructose and, for that matter, all of the monoses. Essentially because mannose, glucose and fructose afforded the same osazone, Fischer utilised this knowledge to fix the configuration at C-2 as (+) for glucose and (−) for mannose.

D-Mannose was also a key substance in establishing, once and for all, the preferred ring structures of the sugars. Because of unusual optical rotatory properties, Hudson had assigned a six-membered ring structure to β-D-mannose and a five-membered ring structure to α-D-mannose. The Haworth school believed that both anomers had a six-membered ring, and resolved the problem by showing that hydrolysis of an oligosaccharide that could only possess a six-membered ring yielded the same compound as occurred naturally. Therefore, α-D-mannose had a six-membered ring in spite of its anomalous optical rotatory properties.

Since that time it has become clear that many of the properties of sugars can be determined by considering the stereochemistry of their puckered six-membered rings. Even reasons for sweetness and varying sweetness are beginning to be revealed by this approach. In the long run, mannose will probably prove to be a key compound in structure–taste studies, as β-D-mannose is bitter, and we do not know why.

STRUCTURE–TASTE RELATIONSHIPS

As a result of studies originating in the USA, and pursued diligently in the UK it is now possible to state that the functional groups for sweet taste in the sugar series are an optimal arrangement of vicinal OH groups, one acting as primary AH in a hydrogen-bonding system and the other as primary B. The former is in one sense a 'proton donor' and the latter a 'proton acceptor'. The initial chemistry of the sweet taste response therefore came to be viewed as a concerted intermolecular interaction between AH, B of a sweet substance and a sterically commensurate AH, B at the receptor site.

The primary AH for aldohexapyranose structures is now known to be OH-4 and B is O-3 (Fig. 1). Galactose, which is only about one-half as sweet as glucose, has primary OH-4 sterically disposed so as to intramolecularly hydrogen-bond the ring oxygen atom (Fig. 2). It cannot, therefore, participate in the required chemistry for the initial taste response of sweetness. The residual sweetness is probably due to subsidiary AH, B units.

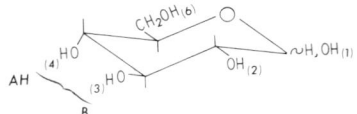 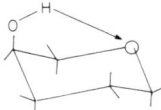

FIG. 1. The D-glucopyranose AH, B unit.

FIG. 2. D-Galactose OH-4 hydrogen-bonded to the ring oxygen atom.

Studies of β-D-fructopyranose also indicated that primary AH is OH-2 and B is O-1 (Fig. 3). However, an impasse is encountered with β-D-fructofuranose, but the impasse is self-inflicted.

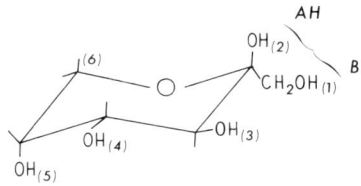

FIG. 3. The β-D-fructopyranose AH, B unit.

Because of the behaviour of fructose in water solution (loss of sweetness during conventional and thermal mutarotation) it was decided that β-D-fructofuranose must be devoid of sweetness. The reason for this appeared to comply with the results for glucose and galactose as OH-2 (primary AH) seemed conveniently placed to intramolecularly hydrogen-bond to O-3 (Fig. 4). In other words, the primary fructose AH unit was effectively eliminated as a participant in the initial concerted intermolecular interaction prerequisite for sweet taste.

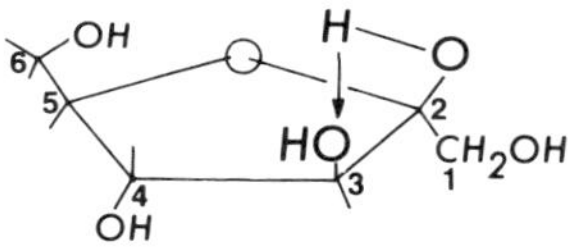

FIG. 4. Proposed hydrogen-bonding of OH-2 to O-3 in β-D-fructofuranose.

However, there had always been reason to be uneasy about that interpretation because of the cloud of uncertainty encompassing fructose chemistry. The uncertainty was due to the fact that only one crystalline form of fructose is known and it is β-D-fructopyranose, with a specific rotation of about $-133°$. Not only was the second major form of fructose in water solution (β-D-fructofuranose) not known in crystalline form but even more unsettling was the fact that its specific rotation was also not known.

As it turns out, the specific rotation of β-D-fructofuranose seems to be about $+78°$.[1] The reason given for the lack of sweetness needs modification and is explained as follows. In Fig. 5, methyl β-D-fructofuranoside (I) is tasteless. Therefore, it can be assumed that OH-2 is somehow hydrogen-bonded intramolecularly in free β-D-fructofuranose (II) because it is also tasteless, although it possesses

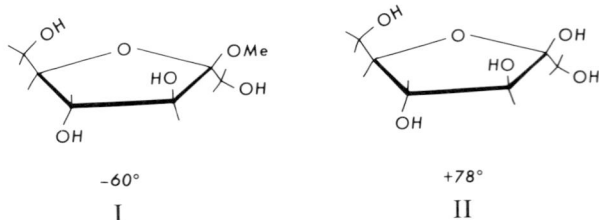

$-60°$ $+78°$

I   II

FIG. 5. Specific rotations and structure of methyl β-D-fructofuranoside and free β-D-fructofuranose.

the very same AH, B unit as does the pyranose modification. In addition, whatever chemistry might be assigned to the properties of OH-2 in β-D-fructofuranose, the fact that substitution of H for CH_3 reverses the chiroptical properties of I must be taken into account. If OH-2 is hydrogen-bonded to OH-3, the taste properties might be affected, but the optical rotatory properties probably would not because OH-2 and OH-3, as a polarisable conformational unit, are nearly symmetrical. Therefore OH-2 must be hydrogen-bonded elsewhere in the molecule.

It was therefore proposed[2] that OH-2 must be bonded to OH-6 across the ring and *this* is the structure of β-D-fructofuranose in water solution (Fig. 6).

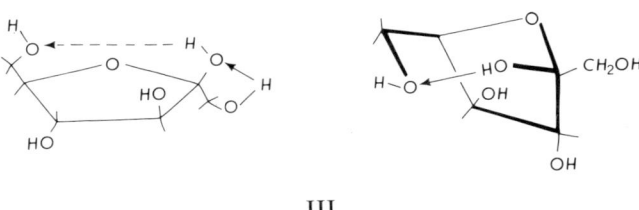

III

FIG. 6. Proposed chelate structure of β-D-fructofuranose in water solution.

Why the compound is tasteless, and also dextrorotatory can now be explained. To do so requires a brief description of an algebraic procedure that permits unambigous identification and specification of the multiple chirality displayed by sugar structures. The method[3] is as follows:

A. General outline of the procedure

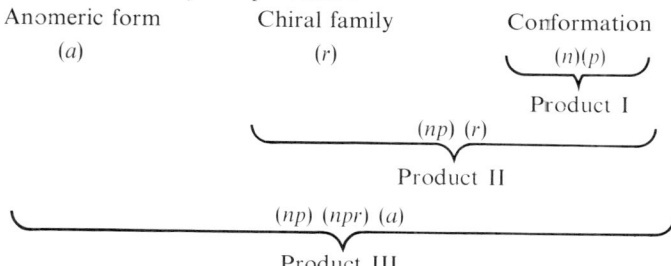

B. Assignment of operands and significance of products
Product I (ring conformation)
Operand n (for numbering). When the carbon atom sequence is clockwise, n is $(+)$; otherwise it is $(-)$.
Operand p (for puckering). When the ring 'O' is up, p is $(+)$; when it is down, p is $(-)$.
Multiply $n \times p$. When the *product* is $(+)$, the conformation is Reeves' C1; when $(-)$, it is 1C.
Product II (chiral family)
Operand r (for reference carbon atom). Observe the bulky substituent at the reference carbon atom. If it is equatorial, r is $(+)$, when it is axial, r is $(-)$.
Multiply $(np) \times r$. When the *product* is $(+)$, the chiral family is D; when it is $(-)$, the family is L. (This is the reason why a conformational specification is ambiguous if the chiral family is not also specified.)
Product III (anomeric form)
Operand a (for anomeric form). Observe the bulky substituent at the anomeric centre. If it is equatorial, a is $(+)$; if it is axial, a is $(-)$.
Multiply $(np)(npr) \times a$. When the *product* is $(+)$, the anomeric form is β-; when it is $(-)$, the anomer is α-. As n^2 and p^2 in the above equation are $+1$, they cancel, and the anomeric form is obtained simply by multiplying $r \times a$. This is why an anomeric specification, which is tied into the configuration at r (Rule 15, rules of carbohydrate nomenclature)[4] is also ambiguous unless the chiral family is also specified.

C. The mnemonic for the calculations
The mnemonic for the significance of the chiral operand assignments and the algebraic products is the most conformationally

favoured aldohexapyranose, which is β-D-glucopyranose in the $^4C_1$ conformation (Fig. 7). All operands are (+), therefore all products are (+).

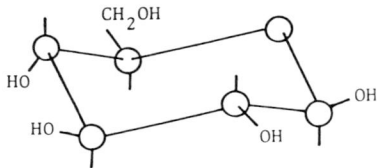

FIG. 7. β-D-Glucopyranose, the mnemonic for calculating multiple chirality specifications.

Using the same type of calculation, but using (+) and (−) signs to signify dextrorotation and laevorotation, respectively, the epimeric rules of Bose and Chatterjee[5] can be employed to ascertain the relative optical rotatory nature, at the sodium D-line, of each chiral centre. In essence, a (−) operand is assigned to a hydroxyl substituent if it lies to the right of the ring oxygen atom or if it lies below the average plane of the ring. For example, a (+) product at the anomeric centre in this case results for an α-D-anomer, and a (−) product for a β-D-anomer. This result is, essentially, an algebraic method of utilising Hudson's rules of isorotation.[6] Applying the dextro–laevo calculation to C-2, C-3, C-4, and assembling the results in one structure yields the compound shown in Fig. 8.

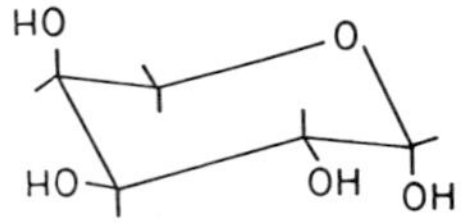

FIG. 8. Computation of the chiral specification of the most dextrorotatory aldopentapyranose.

Chiral feature	n	p	r	a	Product	Result
Conformation	(+)	(+)			(np) = (+)	C1
Chiral family	(+)	(+)	(−)		(np)r = (−)	L
Anomeric form	(+)	(+)	(−)	(−)	(np) (npr)a = (+)	β-

The chiral calculations, when applied to the compound show that it is a β-L-aldopentapyranose in the $^4C_1$ conformation.

Reference to structural tables identifies it as β-L-arabinopyranose and with a recorded specific rotation of +191°, it is the most dextrorotatory of the pentose sugars.

The feature that distinguishes this aldopentose from the others is that the hydroxyl substituents neatly describe a partial left-handed screw pattern with the steepest pitch and the most regular radius. The left-handed screw pattern is required for dextrorotation since solutions of compounds are placed in a polarimeter tube and viewed through the tube in the direction of the light source. Dextrorotatory structures viewed in this way are shown in diagrammatic form in Fig. 9(a).

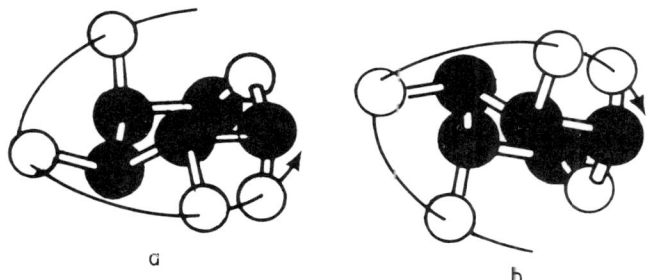

FIG. 9. (a) Dextrorotatory structures; (b) laevorotatory structures.

If carbon atom number 5 is rendered asymmetric by substitution with a $-CH_2OH$ group, it is now (as the chiral calculations will prove) an α-D-aldohexopyranose. In fact, it is α-D-galactopyranose, and with a specific rotation of +151°, it is the most dextrorotatory of the aldohexose sugars.

The most laevorotatory structures in the sugar series must then be the mirror image structures of the above compounds, or β-D-arabinose and α-L-galactose, as shown in Fig. 9(b).

If the β-D-arabinose structure is so substituted at the anomeric centre as to provide an additional $-CH_2OH$ group, then the compound formed is β-D-fructopyranose (laevulose), with a specific rotation of −133°.

With the previous discussion in mind, attention can again be directed toward the intrinsic chemistry of β-D-fructofuranose, and an attempt made to explain simultaneously its lack of sweetness and its optical rotatory properties.

If, instead of bonding H-2 to O-3, it is bonded to O-6 instead, a bicyclic fructofuranose structure in water solution is formed (Fig. 6). The newly formed ring may in itself be dextrorotatory. However, if the polarising power of O-2 should be negated, then the polarisability sequence of the hydroxyl substituents on the furanose ring would be reversed, as shown in Fig. 10. This model simultaneously explains the lack of sweetness for β-D-fructofuranose and also its optical rotatory nature.

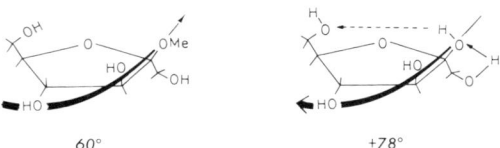

FIG. 10. Proposed reversal of hydroxyl substituent polarisability sequence in transposing from methyl β-D-fructofuranoside to the free furanose sugar.

The problems posed by the intrinsic chemistry of mannose can now be discussed. If the helical pattern of polarisable substituents holds, then mannose has a mixed grouping of polarisable substituents, as shown in Fig. 11. This would explain why Hudson's rules of isorotation do not hold for this compound. As for the intrinsic bitterness of β-D-mannose, it would now seem that this attribute is not sterically directed, in the strict sense, and that a stereochemical transformation in an otherwise sweet molecule that leads to bitterness is more the result of electron redistribution than

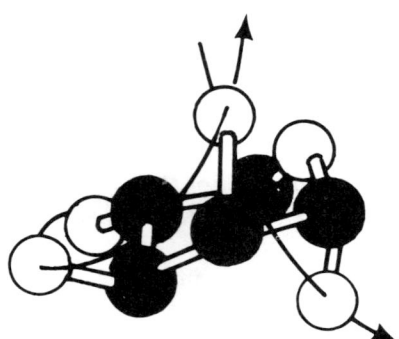

FIG. 11. Mixed grouping of polarisable hydroxyl substituents as found in D-lyxose and D-mannose structures.

it is alteration of development of a steric configuration appropriate for fitting some unknown steric bitter receptor site. For example, the molecule may have been transformed from that of a hard acid to a soft base.

It is with these ideas in mind, and with the employment of the principles described, that the basis is formed for our continued efforts to gain knowledge of the intrinsic chemistry of the sugars or the nutritive sweeteners.

REFERENCES

1. SHALLENBERGER, R. S. (1978). *Pure and Applied Chem.*, **50**, 1409–20.
2. SHALLENBERGER, R. S., BRAVERMAN, S. E. and GUILD, JR., W. E. (1980). *Fd. Chem.*, **5**, 207–16.
3. SHALLENBERGER, R. S., WROLSTAD, R. E. and KERSCHNER, L. E. (1981). *J. Chem. Educ.* (in press).
4. ANON (1963). *J. Org. Chem.*, **28**, 281–91.
5. BOSE, A. K. and CHATTERJEE, B. G. (1958). *J. Org. Chem.*, **23**, 1425–32.
6. HANN, R. M. and RICHTMYER, N. K. (Eds.) (1946). *The Collected Papers of C. S. Hudson*, Vol. 1, Section IV. Academic Press, London, 465–835.

2

Sucrose, the Optimum Sweetener

W. M. NICOL
Tate & Lyle Ltd, Philip Lyle Memorial Research Laboratory,
Reading, Berks, UK

ABSTRACT

The salient issues of recent years are reviewed with respect to their effect on sucrose, which is still by a large margin the prime sweetener in the world. Attitudes to sucrose are changing in the wake of an expanding range of sweeteners. Alternative non-food uses for sucrose are beginning to have a significant effect on supply and demand. Production is, therefore, expected to continue to increase, particularly in view of the favourable energy balance shown by sugar cane and to a lesser extent by sugar beet.

The measurement of sweetness continues to pose multifactoral problems but sucrose remains the standard against which sweeteners are judged.

Technological advances have widened the range of applications of sucrose and this could be an exciting area for food technology. Reference is made to the role of sucrose in the preservation, baking, colouring and flavouring of foods and to conditions for its handling and storage after crystallisation.

The health and nutrition aspects of sucrose are put into perspective by reference to recent pronouncements from leading authorities. Dental caries is still a major area of discussion with evidence which ranges from one extreme to the other in correlation with sucrose consumption. The debate is likely to continue for many years.

INTRODUCTION

'A spoonful of sugar makes the medicine go down'. That old adage epitomises one of the main uses for sucrose—an aid to palatability. After all, man has an innate desire for sweetness which guided him aeons ago to select safe and wholesome foods and, conversely, to avoid those tasting bitter because, as it happens, most toxic plants are in that category.

Since sweetness is generally a pleasurable sensation, when honey, and later sucrose, became available they were used psychologically to increase the palatability of food. Nowadays with an increasing number of alternatives the choice of sweetener is determined principally by cost effectiveness.

Sucrose is the most abundant free sugar in plants and its use has been referred to in documents spanning many centuries. Even in modern times it has been discussed extensively, for example, by Hugill[1] and Nicol,[2] which poses the interesting question—what more is there to say about sucrose? It is an inexhaustible subject because the more we understand the function of sucrose in our food, the more interesting sucrose becomes as an ingredient quite apart from its pre-eminent role as a sweetener.

ECONOMICS

The early stages of a country's growth of GNP are the most favourable for an increase in sucrose consumption because at that stage the need is for food energy, and sucrose very economically stimulates consumption by increasing organoleptic appeal through sweetness.

As the standard of living increases, so does the per capita sugar consumption until a peak is reached at around 50 kg per capita per annum. With further affluence there is then a tendency for the use of sucrose to fall a little as the increased spending power enables a much wider range of alternative foods to be bought.

When consumption is below 15 kg per capita per annum the domestic market predominates while above 40 kg per capita per annum the major outlets are in the food industry. The transition between 15 and 40 kg per capita per annum is seen to be a period for growth in the food industry. Data from the Sugar Year Book[3]

(Table 1) indicate that there is considerable area for growth in sucrose.

TABLE 1
WORLD POPULATION AND SUCROSE CONSUMPTION

	Population (millions) with per capita sugar consumption		
	below 15 kg	15–40 kg	over 40 kg
Europe	0	125	376
North America	0	0	310
Central America	5	28	18
South America	0	103	130
Asia	2116	286	4
Oceania	4	0	18
Africa	260	170	10
	2385	712	866

Not surprisingly then the trend in sucrose consumption is upwards at a rate of about 2% per annum. By 1985 total world production of sucrose for food should reach 100 million tons, 80% of the world increase in sucrose taking place in the developing countries. In real terms sucrose in the UK now costs less than it did say 20 years ago, but consumption is slowly declining due to the availability of alternative sweeteners, principally those derived from starch. These are likely to increase in importance but, for the foreseeable future at least, sucrose will remain the most widely used sweetener. By 1990 it is predicted[4] that in the world sweetener market sucrose, centrifugal and non-centrifugal, will have about 90% of the market and starch-based sweeteners even then only 8%. The pre-eminence of sucrose is due to its universal availability and versatility. McKay[5] catalogued some 40 properties of sucrose which have been summarised in Table 2. They constitute a convenient framework for reviewing the qualities of sucrose.

ENERGY

As the world population expands there is ever-increasing pressure on energy sources for food, heat and power. The fossil fuels are not inexhaustible and it behoves us to make ever-increasing use of

TABLE 2
SALIENT PROPERTIES OF SUCROSE

1. Very desirable sweet taste
2. Bulking agent, diluent and carrier
3. High and ready solubility
4. Good preservative properties through osmosis
5. Cheap, colourless, chemically and microbiologically pure
6. Generates structure through reaction with proteins and starches
7. Nutritious, digestible, non-toxic
8. Rapidly and totally fermentable
9. Forms flavours and colours on heating
10. Good storage stability

renewable sources of energy. The primary source of energy, of course, is the sun whose incident radiation on the atmosphere provides $1.35 \text{ kJ/m}^2/\text{s}$. Plants are able to store energy by virtue of the photosynthetic effect and in terms of re-usable energy the C_4 plants, particularly sugar cane, are the most efficient converters. Being a tropical and sub-tropical plant, sugar cane is able to take advantage of the higher incident solar radiation than is available in the temperate regions.

Austin et al.[6] have calculated meticulously the energy balance of both the sugar cane and sugar beet crops right through to white sugar. The gross biomass energy amounted to approximately 220 GJ/ha/year for beet and 680 GJ/ha/year for cane. However, the efficiency of conversion of incident photosynthetically active radiation into recoverable sucrose was less than 1%. Shore[7] has indicated that in the UK, barley yields a metabolisable energy of 43 GJ/ha whereas sugar beet yields 87.4 GJ/ha as refined sugar plus 34 GJ/ha from the green tops and 35 GJ/ha from the molassed pulp. Table 3 compares beet (UK) and cane (Queensland) energy data.

Using data for sugar beet in California where the support energy is extremely high, the sucrose energy ratio falls to 0.46 which agrees with the value of 0.45 calculated by Rawitscher and Mayer[8] who argued that beet production in the USA should be switched to cereals. Energy conservation may be one thing but the beet growers are more fearful of the economic challenge from corn sweeteners.

While discussing energy it is worth digressing a little to consider the increasing use of sugar cane as a source of power energy by fermentation of its carbohydrate content to ethyl alcohol. Brazil

TABLE 3
ENERGY CONTENT: GJ PER TON SUCROSE[6]

	Beet (UK)	Cane (Queensland)
Sucrose (A)	16·8	16·8
Leaves	10·85	21·75
Bagasse or pulp	6·37	10·23
Filter mud		0·84
Molasses	4·85	2·38
Total crop energy (B)	38·87	52·00
Support energy[a]	23·52	10·32
Transport to consumer	2·5	2·5
Transoceanic transport		1·25
Total energy (C)	26·02	14·07
Maximum energy ratio B/C	1·49	3·70
Sucrose energy ratio A/C	0·65	1·19

[a] Support energy is the total energy input for final products at factory gate and includes fuel, transport, chemicals and the energy equivalent of machinery.

has led the world in commitment because of the ready availability of sugar cane and the pressure on their balance of payments caused by importation of increasing quantities of oil. The forecast for 1980 was a production in Brazil of 6·6 million tons sugar and nearly 4 billion litres alcohol almost all for blending at 15% with petrol (gasoline). By 1990 their production is planned for 20 billion litres alcohol requiring a very great increase in land under sugar cane. The effect of this alternative crop use is expected to lead to a stabilisation of sugar prices in the long term.

SWEETNESS

Sucrose, being the most abundantly available sweetener, is used as the standard reference for sweetness. In its pure state it provides, by definition, only sweetness without any other flavour. Having said that it is clear that all other sweeteners are different, not necessarily better or worse, but different in their physiological profile.

We are all aware that the only equipment we have for measuring sweetness is the tongue and that there is inevitably a statistical variation between tongues. Furthermore, the sweetness of a sweetener is dependent on the conditions of test such as temperature, concentration and acidity. Hyvönen et al.[9] reported that at 5°C sucrose sweetness was reduced by about 10% and at 50°C it was conversely enhanced compared to room temperature (22°C) (Fig. 1).

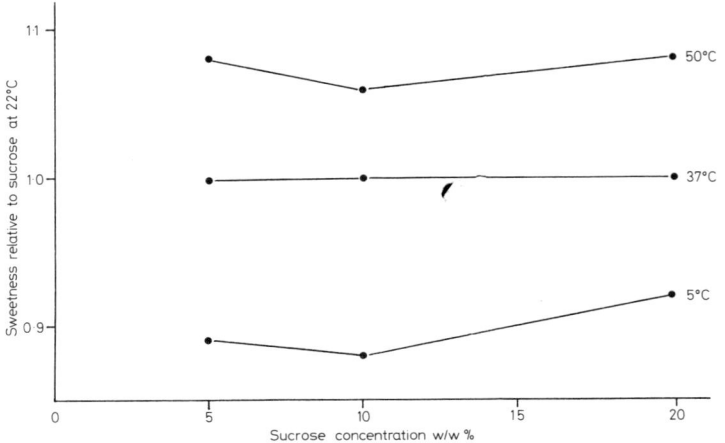

FIG. 1. The influence of temperature on the relative sweetness of sucrose over a concentration range.

The effect of acidity seems more controversial because of the difficulty of comparing sucrose solutions with sucrose–acid solutions. At the supra-threshold levels of sucrose and acids Pangborn[10] showed that citric, tartaric, acetic and lactic acids all reduced the sweetness of sucrose to some extent. Earlier work, on the other hand, had indicated that acetic acid had no effect while lactic, citric and tartaric acids actually increased the sweetness of sucrose. Part of the problem may be in attempting to compare basically dissimilar systems.

The most common sweetness intensity in sweetened foods corresponds to a sucrose solution of between 5 and 12%. Concentrations much in excess of that, although probably preferred, cause much more rapid satiety as is evidenced by the fact that those who consume high concentrations of sucrose, as in confectionery, do not necessarily consume more sugar calories. In a study[11] using Moskowitz's magnitude estimation scaling method which claims to shorten dramatically the time and cost of developing consumer-attractive products, it was shown that cola drinks were preferred at a much higher sweetness level than is commercially provided (Fig. 2).

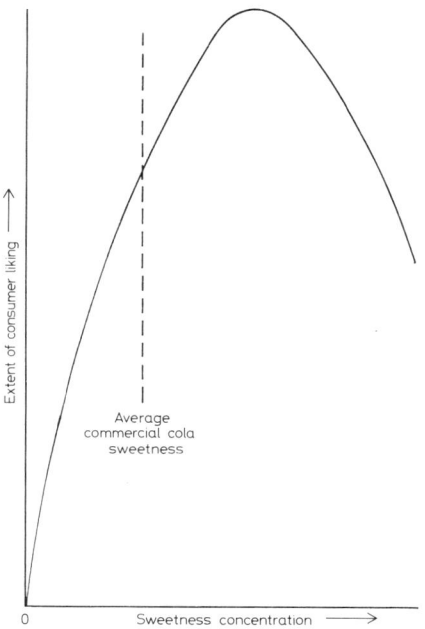

FIG. 2. Consumer preference and sweetness level in cola drinks, according to Moskowitz's scaling method.[11]

The difficulties are great enough in establishing the commonly accepted standard of sweetness but when assessing the sweetness of other sweeteners relative to sucrose the problem is much more complex. Apart from the possible variability of taste panels, there is a clear trend that relative sweetness depends on the sweetener, its concentration, the temperature and the food environment in which it is being tasted. Hyvönen et al.[9,12] have reviewed varying relative sweetness with respect to temperature, concentration and acidity for a number of sweeteners and clearly indicate the general lack of agreement due to the multifactorality of the dependence.

Oral preconditioning can also influence the assessment of sweetness. Thus Thompson and Campbell[13] reported the enhancement of hedonic rating of sucrose solutions if there is prior conditioning of the oral cavity with 2-deoxy-D-glucose.

In conclusion, the measurement of sweetness is not easy when even the perception of the standard, sucrose, is influenced by so many factors.

BULK

Apart from contributing sweetness to foods, sucrose fulfils another major role acting as a filler, a diluent and, at times, as a carrier for trace ingredients, typified by flavours and colours. For over 100 years, since the invention of the vacuum crystalliser and the centrifugal machine (for separating crystal from mother liquor), sugar has been most widely available in its macrocrystalline form where the crystals have been upwards of $300\,\mu m$ in size. That the product has remained virtually unchanged during the period indicates an element of true adequacy.

However, recent developments in processing have enabled sucrose to be produced efficiently and continuously as novel microcrystalline agglomerates; this has many potential advantages. These products, known as transformed sugar, have been produced in several tropical countries for over 100 years and, more recently, developed in Japan and known as Q sugar. With major advances in the technique to improve the control of the transformation process[14] we should see more of the product in the developed world. Transformed sugar has a large absorptive surface area, low bulk density and faster solubility. The extensive surface area enables, for example, brown sugars to be produced, which remain free-flowing for a very long period of time. After 3 years in a warehouse such a brown sugar was still pourable. Clearly, such a characteristic is ideally suited to bulk-handling in the food industry. Its more rapid solubility has so many potential applications that it would take too long to list them all, but where cold dissolution is required, as in powdered drinks, it seems to meet a need. The increased surface area enables the sugar to be used as a carrier for colour, flavour and other micro-ingredients. A technical variation in production enables it to be used very readily in compression tabletting which offers opportunities both in the confectionery and pharmaceutical industries.

Pursuing a different line, Niedick and Babernics[15] demonstrated that freeze-dried sucrose in its amorphous state had several orders-of-magnitude greater absorptive capacity for aroma compounds, than crystal sucrose. However, the commercial cost of such a procedure would limit it to applications of high value only. Besides, amorphous sugar is considerably more deliquescent than the crystalline form as can be deduced from the moisture equilibrium isotherms.[2]

The advantage of sucrose as a bulking agent is its high solubility, about 2 g per g in water at ambient temperature, and its reversibility of crystallisation and solution as used frequently in the confectionery industry. With suitable crystallisation inhibitors, such as glucose, a relatively stable sucrose glass, the boiled sweet, can be made with as little as a few percent water. There was considerable interest a few years ago in an additive that would, on trace addition to sucrose solutions, inhibit crystallisation altogether. Several candidates were identified but the cost of regulatory approval for an uncertain market return ensured that these stayed in the laboratory.

More and more computer control of processing is being used and tabular basic data are being translated into mathematical equations. For such an application Genotelle[16] has expressed viscosity in terms of a function(ϕ) of molar concentration(N) and temperature(t):

$$\log (\text{viscosity}) = 22 \cdot 46 N - 2 \cdot 114 + \varphi(t) (1 \cdot 1 + 43 \cdot 1 \ AN^{1 \cdot 25})$$

By the use of the appropriate weighting factor(A) the relation can be extended to impure solutions.

PRESERVATION

The function of sucrose as a preservative relies mainly on its ability to bind water and so make it unavailable to support spoilage organisms. For instance, few fruits contain sufficient sugar to ensure the preservation of their flavours and textures so the addition of sucrose in fruit processing has been customary and necessary. The most obvious application must surely be in jam and preserves. This is a contracting sector of the market but the frozen food sector is rapidly expanding. The addition of sucrose prior to quick freezing inhibits oxidative degradation in the fruit. Going a stage further, in canning, sucrose diffuses into the fruit cells to preserve the texture and overall appearance while minimising oxidation when ultimately exposed to the atmosphere. Prozucek *et al.*[17] showed inhibition of enzymes in apple puree by sucrose. Ascorbate oxidase was fully inhibited by 30% sucrose while catalase required 50% sucrose.

In quite a surprising and clearly extravagant experiment,[18] a 25% sucrose solution was sprayed on lime trees in blossom. The effect was a doubling of fruit set, a reduced fruit drop, delayed maturity and a significant increase in fruit size and ascorbic acid content.

However, the use of sucrose as a preservative is not confined to the obviously sweet type of product: it is an important ingredient in meat curing.[19] In fermented sausages sucrose acts as a substrate for acid formation which inhibits production of toxins by food-poisoning bacteria. About 1% sucrose is the normal level of addition which is below the sweet threshold yet allows a mellowing of the harshness caused by the added salts. It assists in optimising consumer acceptability.[20] In this respect sucrose also reduces the losses from shrinkage during processing and cooking, and at the same time maintains a juicier product on the plate.

PURITY

Sucrose is known as being pure, white and healthy, and not without good reason for it is one of the most chemically pure foods available in the world. The standard ubiquitous granulated sugar has a dry impurity of less than 0.1%. It is substantially free of any organisms or toxins that would lead to a health hazard. Included in this context must be the clarity of aqueous solutions of sucrose which makes it such a useful bodying agent in soft drinks. Table 4 shows the relative purity of comparative food ingredients.

TABLE 4
PURITY OF FOOD INGREDIENTS

Food ingredient	Minimum purity (%) (Food Chemicals Codex)
Acetic acid	99.5
Ascorbic acid	99
Sodium bicarbonate	99
Citric acid	99.5
Sodium chloride	99.5
Sorbitol	91
Sucrose	99.7[a]

[a] Codex Alimentarius

COST

When assessing the cost of sucrose relative to other foods, it is difficult to fix a satisfactory basis for comparison, for prices change according to supply and demand, political intervention, quality and location, to mention but a few, albeit major, factors. As was explained above, the relative sweetness of other sweeteners is variable. However, with the assumptions stated, the relative cost per unit of sweetness on tonnage deliveries is shown in Table 5. At the domestic level, the comparative energy costs are shown in Table 6.

TABLE 5
RELATIVE COST AND ENERGY VALUES OF SEVERAL SWEETENERS

	Sweetness factor	Calorie value per unit of sweetness	Relative cost per unit of sweetness
Sucrose, crystalline	1	4	330
Glucose, liquid solids	0.7	5.7	400
Isoglucose, liquid solids	1.0	4	310
Fructose, crystalline	1.2	3.3	1000
Sorbitol, powder	0.5	8	1500

TABLE 6
COST OF ENERGY FROM VARIOUS FOODS

	Cost (pence/100 kcal)
Sucrose	1·0
Flour	0·9
White bread	1·9
Potato	2·1
Rice	2·5
Spaghetti	2·1
Beefburger	8·3

STRUCTURE

Sucrose acts as a regulatory factor in reactions at all stages of baking but especially in crumb formation. As the sucrose level is decreased the crumb becomes thinner and elastic, and less acceptable.

Using differential scanning calorimetry, Wootten and Bamunuarachchi[21] studied the effect of sucrose on wheat starch gelatinisation. It appeared that the temperatures at the onset and conclusion of the effect were not changed by sucrose concentration but that the temperature at which the effect reached a maximum was higher. This ties in with the earlier work. The effect is nearly proportional to the concentration of sucrose in solution indicating that it is partly due to competitive water binding. Bean et al.[22] studied the effect of sugar solutions up to 60% on the gelatinisation temperature of wheat starch. At a given sugar concentration, the gelatinisation temperature was higher for sucrose than for glucose and fructose. In a sugar to flour mix of 1·3:1, the optimum water ratio was 1 for sucrose, 0·85 for fructose and 0·65 for glucose. These water levels seemed to reflect the need to achieve the same starch gelatinisation temperature. Practically, the most effective starch gelatinisation temperature for cakes is approximately 90°C corresponding to a sucrose:water ratio of 1·3 in the batter.

Sucrose increases the temperature of denaturation of proteins. The effect is similar to gelatinisation in being related to water binding.[23] Mostly, it is manifested in the relatively rigid structure of a cake confection. Less rigid products are also possible as reported by Razanajatovo et al.[24] who studied the gel formation of combinations of whey protein and sucrose when heated. A combination of 30% sucrose and 5% whey protein maintained at 112°C for 5 min gave a smooth homogeneous gel resembling egg custard.

So far the effect on texture has arisen from sucrose in solution, but crystalline sucrose also has a part to play. The most obvious example is the meringue. Fine grained sugar, because of its greater surface area, provides more stabilisation of egg white. It is effectively a filler which gives solid structure to the protein foam, so the finer the crystal the finer the texture.

In pastry, coarse crystals around 800 μm in size yield large pores. The pastry also tastes sweeter because the crystals do not dissolve in the baking process.[25] These crystals and their size determine the product pore size and distribution and so the appearance, softness and structure.

NUTRITION

Many sources report that sucrose, being pure, white and sticky, is the optimum sweetener for dental caries. Others claim that honey is more cariogenic than sucrose, that bananas, grapes and raisins are more so than caramels, bread and raspberry jam. While that may be so, it is impractical, for a variety of reasons, to stop consuming sugars any more than we should stop going out in the rain lest we get wet. Caries requires:

1. Teeth that are susceptible to decay.
2. Bacteria that cause decay.
3. Foods that when acted on by bacteria form acids.

Susceptibility seems to relate to fluoride concentration in the teeth, and that is why fluoridation is such a successful means of reducing caries. One part per million in the water supply reduces the incidence of caries by 50%. Heredity also plays an important part in determining susceptibility of teeth to decay. The most virulent of the bacteria causing tooth decay is *Streptococcus mutans*. Prevention of infection by it is thought to produce an effective reduction in caries but is easier said than done. Vaccines are being studied that would produce a secretory immunoglobulin antibody, but there is a long way to go. Alfano[26] believes that an iron deficiency may be an important factor leading to caries. While most people believe that sucrose is the main cause of caries, it has to be admitted[27] that currently no adequate procedure is available for the evaluation of the caries-producing ability of an individual food.

Carbohydrates, and sucrose in particular, are substrates for *S.mutans* which reduce the pH in the mouth rendering the teeth more susceptible, and simultaneously generate polysaccharides which adhere to the teeth allowing a sheltered microenvironment for the acids to etch the teeth. Thus the practical course of action is not to stop eating but to practise good oral hygiene by brushing the teeth free of polysaccharide as soon as possible after eating.

Probably the most striking example of dental progress can be seen by considering three generations of one's own family. It will be surprising if there is not a very marked difference in oral condition of present children from that of their parents and grandparents at the same age. A recent review[28] sums up the present status and gives a clear picture of how progress is being made.

Even if it can be demonstrated that sucrose causes caries in laboratory animals, epidemiological studies on humans give anything but a clear picture. Hargreaves[29] at Harvard made a study of several hundred children in four communities in Ontario covering a range of rural through to industrial areas. Very detailed examination of their teeth and the actual food consumed enabled a variety of correlations to be established. The mean daily intake of sugars was 150 ± 54 g and of that 83 g was sucrose. The correlation factor relating sucrose intake with dental decay was 0·048 (Fig. 3) hardly a convincing correlation. An international advisory group appointed by several national nutrition foundations has just published a report[30] underlining the need for a great deal more research in nutrition, dental caries and oral health.

Kroyer and Washuttl[31] showed that sucrose has no effect on essential amino acids nor on water-soluble vitamins except B and C which experience only a small loss. In the same paper the effects of sodium saccharin and sodium cyclamate were also reported showing, in some cases, very serious destruction of nutrients.

Sucrose is a significant source of calories in the diet of most developed countries. It contributes to the satiety and organoleptic values of food. In spite of speculation that sucrose is a major contributory factor in the aetiology of numerous diseases, the weight of evident disproves the case. The Federation of American Societies for Experimental Biology, in expressing a neutral assess-

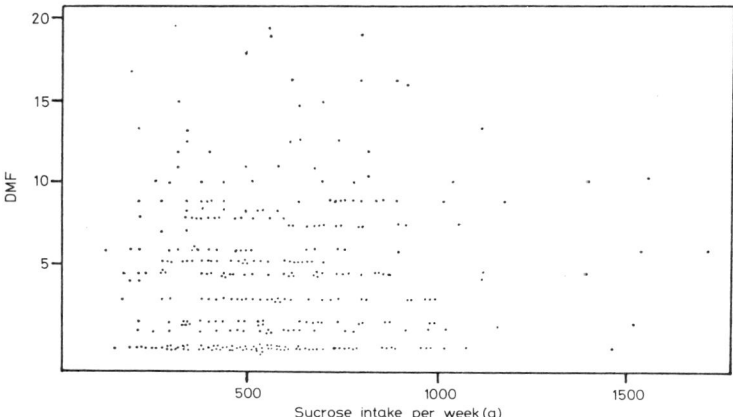

FIG. 3. Results of a study relating to the incidence of dental caries (as decayed, missing or filled) in children and the level of dietary sucrose intake in the form of a scattergram.

ment, said 'other than the contribution made to dental caries, there is no clear evidence in the available information on sucrose that demonstrates a hazard to the public when used at the levels which are now current and in the manner practised'. The adequacy of a diet is determined by the balance of foods selected, any one of which on its own could be considered detrimental to the health of the body if consumed in excess. There is no substantive evidence that sucrose-containing foods are worse than any other foods.

COLOUR AND FLAVOUR

Both colour and flavour arise from the thermal degradation and caramelisation of sucrose. There are two main types of caramel. Colouring caramel is obtained by heating sucrose in the presence of catalysts either in an open pan or continuously under pressure at a temperature in excess of 500°C for a very short time. Flavour caramel demands less vigorous thermal treatment to avoid the generation of bitter compounds.

Brown sugars, because of their coating of syrup containing natural and thermally generated impurities, have specific flavour characteristics. Hassett[32] describes some of these well-known brown sugars including demerara and muscovado. One of the difficulties associated with such sugars is the maintenance of adequate quality control in the producer countries and in shipping.

Just as important is the use of the flavour-enhancing properties of sucrose when used at levels either above or below the threshold of sweetness. Being pure, white and tasty, it has long been used for increasing the acceptability of unpalatable food by using sweetness to mask unpleasant flavours. Enhancement is a more subtle psychological tool but it is used very effectively in this way in a wide range of foods, such as chocolate, cereals and fruit products. Below the threshold of sweetness at levels of addition up to 2%, it is used to improve sauces, soups, mayonnaise, cheese and yogurt amongst many others. Everson[33] reports on the use of sucrose in cured meats and how, through flavour intensification, it affects the acceptability of the product. Sausages are beneficially improved with about 1% sucrose added. Sucrose also helps to achieve another objective of curing—the stabilisation of the red colour of meat which is deemed to be highly desirable by the consumer.

STORAGE

As is well known, sucrose is an ideal source of energy both for humans and microorganisms. The latter need moisture to survive, so for prolonged storage sucrose has to be kept dry. Raw sugar crystals have a thin film of mother liquor surrounding them. Although this is normally a saturated solution, with a low water activity (0·86), some organisms can still survive. Under adverse conditions the film can become more dilute adding to the microbiological lability. It is not difficult to see that raw sugar is more of a convenient commercial commodity rather than a stable food ingredient.

Hence there is a need to purify sucrose to the pure, white and handy form that we know so well. The pre-eminence of sucrose must be due to the ease with which it can be produced in the dry crystalline form, and its tolerance of a wide range of humidities. It equilibrates at a virtually constant moisture level over the humidity range 10 to 85% (Fig. 4). It can withstand wide fluctuations in ambient temperature provided that the RH does not exceed 85%, for above that the surface of the crystal slowly becomes sticky due

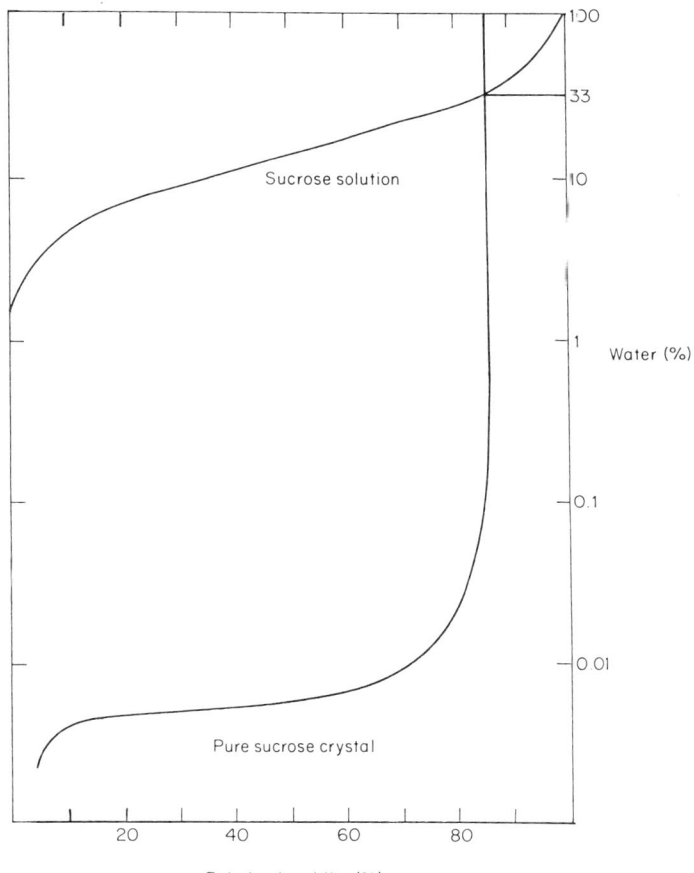

FIG. 4. Relative humidity isotherms for sucrose crystals and solutions.

to the adsorption of water. Cycling humidity above and below the 85% level ultimately leads to caking.

On separating the crystals from the mother liquor the residual film on the crystals is rapidly dried to a glass from which the water diffuses gradually as the glass crystallises. This water leaves the crystal in a matter of hours. However, in the process of crystallisation a small quantity of water is trapped within the crystal itself and diffuses out over a period of about 14 days. It may be thought that this is splitting hairs when examined in absolute terms for the moisture level will be halved over the 14 days from initial crystallisation, but unless the water is allowed to escape it can cause serious problems. Temperature gradients in storage will also cause moisture to migrate to the colder regions. Freshly crystallised sucrose sealed in a plastic bag, or other moisture barrier, will generate sufficient free water to make the sugar creepy and susceptible to spoilage. Sucrose for bulk delivery is, therefore, carefully conditioned to minimise potential trouble in a customers' bulk store.

CONCLUSION

Sucrose, whether produced from cane or beet, is a valuable economic crop that will continue to play a leading role in feeding the ever-expanding world population. The versatility of its properties will ensure that it will have diverse uses in making food more attractive.

REFERENCES

1. HUGILL, A. (1979). In: *Developments in Sweeteners—1*, Eds. Hough, C. A. M., Parker, K. J. and Vlitos, A. J. Applied Science Publishers, London, 1–42.
2. NICOL, W. M. (1979). In: *Sugar: Science and Technology*, Eds. Birch, G. G. and Parker, K. J. Applied Science Publishers, London, 211–30.
3. INTERNATIONAL SUGAR ORGANIZATION, (1979). *Sugar Year Book*, International Sugar Organization, London, Table 1.
4. EARLEY, T. C. (1980). *The Future of Sugar*, Financial Times Conference Organization, London, 62–9.
5. MCKAY, D. A. M. (1979). In: *Health and Sugar Substitutes*, Ed. Guggenheim, B. S. Karger, Basel, 322.
6. AUSTIN, R. B., KINGSTON, G., LONGDEN, P. C. and DONOVAN, P. A. (1978). *J. Agric. Sci., Camb.*, **91**, 667–75.
7. SHORE, M. (1978). *J. Agric. Sci., Camb.*, **91**, 674.
8. RAWITSCHER, M. and MAYER, J. (1979). *Food Policy*, **4**(2), 138–9.
9. HYVÖNEN, L., KURKELA, R., KOIVISTOINEN, P. and MERIMAA, P. (1977). *Lebensm.-Wiss. u Technol.*, **10**, 316–20.
10. PANGBORN, R. M. (1965). In: *Proc. 1st Internat. Congr. Food Sci. Technol.*, Vol. III, Ed. Leitch, J. M. Gordon and Breach Science Publishers, New York, 291.
11. ELIAS, S. (1980). *Food Engineering*, (September), 102.

12. HYVÖNEN, L., KURKELA, R., KOIVISTOINEN, P. and ALA-KULJU, M. (1978). *Lebensm.-Wiss. u Technol.*, **11**, 11–14.
13. THOMPSON, D. A. and CAMPBELL, R. G. (1977). *Science*, USA, **198**(4321), 1065–8.
14. NICOL, W. M. (1975). *Brit. P.* 1 460 614.
15. NIEDICK, E. A. and BABERNICS, L. (1979). *Gordian*, **79**(2), 35–44.
16. GENOTELLE, J. (1978). *Ind. Alim. Agric.*, **95**, 747–55.
17. PORZUCEK, H., WENTYKIER, J. and HORUBALA, A. (1978). *Przemysl Spozywczy*, **32**(3), 112–3.
18. RANJIT, K. and SINGH, J. P. (1978). *Agric. Agro Industries J.*, **11**(10), 15–16.
19. LACHMANN, A. (1975). *The Role of Sucrose in Foods*, International Sugar Research Foundation Inc., Bethesda, 237–54.
20. PYRCZ, J. and PEZACKI, W. (1978). *Proc. Eur. Meeting Meat Research Workers*. No. 24, F2:1–F2:6.
21. WOOTTON, M. and BAMUNUARACHCHI, A. (1980). *Stärke*, **32**(4), 126–9.
22. BEAN, M. M., YAMAZAKI, W. T. and DONELSON, D. H. (1978). *Cereal Chem.*, **55**(6), 936–52.
23. BACK, J. F., OAKENFULL, D. and SMITH, M. B. (1979). *Biochem.*, **18**(23), 5191–6.
24. RAZANAJATOVO, L., ALAIS, C. and PAUL, R. (1978). *Lait*, **58**(578), 483–95.
25. TRACHSEL, B. (1979). *Zucker und Susswaren Wirt.*, **32**(6), 207–8.
26. ALFANO, M. C. (1980). *Food Technol.*, **34**(1), 70–4.
27. ANON. (1978). *Nutr. Rev.*, **36**(8), 249–51.
28. SANDERS, H. J. (1980). *Chem. & Engnr. News*, (February 25), 30–42.
29. HARGREAVES, J. A. (1980). Sucrose and total sugar intake and dental caries of Ontario children. Presented at Kellogg Nutrition Symposium, Toronto, Ontario, 7–18 March.
30. *The International Advisory Group Report on the Relationship between Diet, Nutrition and Dental Caries*, (1980). Nutrition Foundation, New York.
31. KROYER, G. and WASHUTTL, J. (1979). *Z. Ernahrungswissenschaft*, **18**(2), 139–44.
32. HASSETT, J. (1979). *Bakers Rev.*, (March 21), 23.
33. EVERSON, C. W. (1978). *Meat Industry*, (August).

3

The Use of Sugars in Confectionery

P. H. WIGGALL
International Scientific Standards,
Cadbury Schweppes Ltd,
Bournville, Birmingham, UK

ABSTRACT

Following a brief classification of different types of confectionery and their position in the market, the main types of carbohydrate nutritive sweeteners used in their manufacture are considered. The ability of carbohydrate nutritive sweeteners to provide not only sweetness, but a variety of textural properties through their ability to form 'glasses', be crystallised in a controlled manner, act as osmotic preservatives, enhance or produce reaction flavours, provide bulk and calories, etc., are outlined, and their uses in different types of confections are described. The extent to which different sugars can be used as alternatives to sucrose, not only from practical considerations but also in the case of chocolate products from a legislative standpoint, is briefly discussed in relation to EEC requirements.

INTRODUCTION

The use of the term 'confectionery' is associated very often with a number of types of products. These include flour-based confections (such as cakes, shortbread, pastries, tarts, biscuits, macaroons and similar products), and chocolate and sugar products more popularly known as 'sweets'. Although sugar plays an important part in many flour confectionery products, this paper is confined to the use of sugars in chocolate and sugar confectionery types of products.

Historically, within the UK at least, sugar confectionery manufacture began much earlier than chocolate manufacture, and started with the 16th and 17th Century apothecaries who used to coat their pills with sugar to render them more palatable. Gradually they found that there was a demand for sweet confections without the active medicinal ingredient whose taste the sugar-flavoured coating was intended to hide. Subsequently firms specialising in such formulations became established. Chocolate production did not begin until the 19th Century, although the consumption of a cocoa-based drink started much earlier, the first chocolate house opening in Bishopsgate in 1657.[1]

The present day industry is a sophisticated one and a major UK exporter. The level of consumption has remained fairly constant in the UK for many years at about 11·9 kg per annum per person or approximately 8 oz (225 g) per week.[2] Sugar in the form of sucrose is of course a major ingredient of confectionery but it is incorrect to think of confectionery as the main source of sucrose within the diet. Only some 13% of sucrose consumed in the UK is consumed as confectionery, although the proportion of the total consumption of glucose syrup, made by partial hydrolysis of starch, which is consumed as confectionery, is higher, reaching some 30%. The consumption of chocolate-based confectionery and of sugar confectionery within the UK are approximately equal.[2]

Legislative Categories

Within the UK there are currently three legally recognised groups of such products: chocolate products, sugar confectionery and chocolate confectionery.[3,4] Those known as chocolate products include: milk and plain chocolate blocks and bars, filled blocks or enrobed products with more than 25% of a legally defined type of chocolate (many of which fall within the category known in the trade as 'count lines', as they are often sold by number), and assortment items, such as boxes of chocolates, in which more than 25% of a true chocolate has been used. These items are subject to detailed compositional and labelling requirements which stem from the EEC directive on cocoa and chocolate products and subsequent amendments.[5]

The second major group of products are the sugar confectionery items, of which the characterising ingredient is a sweetening carbohydrate. These embrace a great variety of types of products such as boiled sweets, toffees, fudges, fruit chews, nougats, marshmallows, candy, liquorice, gel-type products such as Turkish delight and tabletted products, etc.

Finally there is an intermediate category in legal terms but which may be considered to bridge the gap between chocolate products and sugar confectionery. These are the chocolate confectionery products, which often consist of a predominant sugar confectionery centre with less than 25% by weight of a true chocolate coating. Although they are characterised by chocolate, because the level present is less than 25% there is insufficient for them to be considered as filled chocolate products in the terms of the EEC directive[5] and consequential legislation in the various EEC member states.[4] Imitation chocolate products, usually made by using a fat other than cocoa butter in a chocolate-type formulation at levels in excess of 5%, which is permissible in some EEC countries when labelled so as to make it clear that it is not true chocolate, may also be considered here.

Although these legal distinctions exist, in practice confectionery spans a continuous spectrum of products. This is made possible by virtue of the varied properties of the nutritive carbohydrate sweeteners available which the confectioner, through his skill and imagination utilises in conjunction with a number of other major ingredients, such as milk, fats, dried fruit and nuts, to formulate the products.

Confectionery, therefore, is made by compounding ordinary foods like sugar, milk or fats into a variety of attractive food products.

SUGARS USED IN CONFECTIONERY MANUFACTURE

A variety of different nutritive carbohydrate sweeteners are used in the manufacture of confectionery, the most significant of which are sucrose, various glucose syrups, lactose (primarily through the incorporation of milk or milk products), invert sugar, and to a lesser extent dextrose and fructose as individual sugars, and sugar alcohols such as sorbitol. Although it has been mentioned that various glucose syrups are used, detailed consideration will not be given to the many kinds of specialised products available, as these have been the subject of a complete symposium within this series.[6] Suffice it to say that a great variety of products are available, produced by partial hydrolysis either with mineral acid and/or various enzymes. These products have varying dextrose equivalents and carbohydrate profiles and fulfil a variety of specialist needs. Hydrogenated glucose syrups such as Lycasin® may also be used for certain specialised applications but are not currently used in large quantities in the UK.

FUNCTIONS OF NUTRITIVE SWEETENERS IN CONFECTIONERY

It is often popularly thought that the main reason for the inclusion of sugars in confectionery is to provide sweetness. However, sweetness is only one of many attributes of sucrose and the various other nutritive sweeteners. Indeed in some situations the sweetness of sugars can be a positive disadvantage when using some of their other properties in formulating confections.

Apart from providing sweetness sugars perform a variety of other functions, supplying bulk and calories (most are considered to contribute about 16 kJ (3·75 kcal) per g),[3] providing a variety of textural attributes through their ability to form glasses or to be crystallised in a controlled manner, and acting, through their solubility, as osmotic preservatives. Furthermore, the elevation in boiling point that arises by virtue of their solubility also means that high temperatures can be reached during the production of products such as boiled sweets, which is of benefit in producing microbiologically safe products. The viscosity characteristics of solutions of nutritive sweeteners may also be of relevance in providing a pleasant mouthfeel on consumption, which may be further enhanced in some cases by the cooling effect noticed as a result of the negative heat of solution, characteristic of sugars such as dextrose. They may also provide a 'clean tasting' background and enhance the flavour of other ingredients such as fruits. In other confections reducing sugars, through reaction with other food ingredients, may be involved in browning or Maillard-type reactions during caramelisation, providing interesting flavours in their own right. Their ability to be fermented to produce alcohol and, through their solubility characteristics, to lower freezing point, is not used in confectionery manufacture although these properties

® Registered Trade Mark of Roquette Frères, Lestrem, France.

are obviously of paramount importance in other industries such as those concerned with the production of alcoholic beverages and ice cream.

Some nutritive carbohydrate sweeteners have a greater affinity for water than others and therefore they can be of varying hygroscopicity. This property can be a positive advantage in some situations in helping to retain moisture and thus prevent the confection drying out, but may be a positive disadvantage in other situations causing deterioration or processing difficulties.

Thus it can be seen that nutritive carbohydrate sweeteners may by virtue of their varying properties fulfil many functions in confectionery (and indeed in other foods also) as well as providing sweetness. These are further considered in relation to a variety of different types of confectionery in the following sections.

Chocolate

The main nutritive sweeteners present in milk chocolate are normally sucrose and lactose. Sucrose may be present up to 55% by virtue of the EEC directive on cocoa and chocolate products and consequential national legislation,[4,5] although typically it is present at about 45%. Lactose is typically present at about 9% in UK-type milk chocolate and arises through the incorporation of the milk. Consumers in many European countries prefer a milk chocolate which is less milky in character and which in consequence contains less milk solids and thus less lactose.

Glucose syrups are not normally used in the manufacture of chocolate and indeed can create great problems in processing if present in more than small amounts as they cause an increase in the viscosity of the molten chocolate. Although the viscosity may be reduced either through the use of additional cocoa butter or by the use of emulsifiers, other processing problems may occur. In plain chocolate, which does not of course contain milk solids, lactose is not normally included, the only sugar normally used being sucrose, which typically may be present at about 50%. Sorbitol may be used in formulations for diabetics, and chocolate made with fructose also for diabetics was at one time available in Germany.

It is permissible within the EEC to use other sugars in chocolate as well as sucrose. Dextrose, fructose, lactose (incorporated as the sugar as opposed to arising from the milk) and maltose are specifically permitted up to 5% of the weight of the product without declaration. Dextrose may be incorporated at levels between 5 and 20%, in which case the name of the product has to be accompanied by a declaration of its presence.[4, 5]

Methods of manufacture

There are two main methods of manufacture of milk chocolate. One method involves initial removal of the water from the milk to produce milk powder, which is then mixed with the cocoa mass (i.e. roasted ground cocoa nibs), sugar and additional cocoa butter, prior to subsequent grinding and further processing; whereas the alternative process, known in the industry as the crumb process and widely used in the UK, does not involve the initial isolation of milk powder. Instead full cream sweetened condensed milk is first produced. This is then subsequently mixed with the appropriate quantity of cocoa mass and dried under vacuum to about 90% solids. The mixture is kneaded to a paste which both stimulates crystallisation and limits the growth of the crystals, and evaporated to yield a stable intermediate product known as milk chocolate crumb, which can be stored by virtue of the low moisture content. It contains all of the sugar and lactose that will be present in the finished milk chocolate but does not include all the fats such as cocoa butter or other vegetable fats which are permitted in some countries, and which may be incorporated in subsequent processing. The ingredients, including reducing sugars such as lactose, milk protein and cocoa constituents, are dried together to produce the crumb and as a result desirable flavours can be developed.

The size of the particles within the finished chocolate, including the size to which the sugars present are ground, can have a significant effect on the mouthfeel and texture of the product. If the particle size distribution is too coarse the chocolate will tend to be perceived as gritty, but if too fine may similarly not be appreciated by all consumers. Chocolate sold within different markets may be produced to different particle size criteria depending on consumer preference.

Boiled Sweets

Boiled sweets, sometimes referred to as 'high boilings', make use of the property of sugars in suitable combinations to form a 'glass'. They are in fact highly viscous supersaturated solutions of sugar produced by boiling to yield a product of low residual moisture content (about 3%). They are usually made using a mixture of sucrose and 42 DE glucose syrup in proportions normally between 1·2:1 and 1·5:1 on a solids basis, and of course may include a variety of flavourings and permitted food colours. In view of their high soluble solids content they have a good potential shelf life with respect to microbiological spoilage. If invert sugar is used they have a greater tendency to deteriorate by becoming 'sticky' through being hygroscopic. When this occurs there is a reduction in the viscosity of the outer part of the sweet allowing crystallisation of the sucrose to occur. It is also possible to make boiled sweets from other types of carbohydrate sweeteners such as Lycasin, a hydrogenated glucose syrup, but most are currently formulated using sucrose and glucose syrup. Considerable work on the conditions needed to prevent crystallisation has been carried out at the Food Research Association at Leatherhead, England.

Fondants

When making fondants the confectioner obviously uses the crystallisation behaviour of sugars, particularly of sucrose and dextrose. Although confectionery products containing very high concentrations of sucrose are likely to crystallise, a property used in making fondants, in order to provide adequate shelf life it is necessary to make sure that the total soluble solids concentration is sufficient to inhibit microbiological spoilage. At 20°C a pure sucrose solution is saturated at a concentration of 67·7% w/w[7] but through the inclusion of either glucose syrup or invert sugar, the total soluble solids concentration can be raised to produce a product of sufficient osmotic potential to inhibit microbiological spoilage. Dextrose is less soluble than sucrose but mixtures of dextrose and sucrose can be produced with a greater total soluble solids concentration than that of the individual sugars, although the solubility of each is depressed. Lees and Jackson[8] have stated that a syrup solids content of 84% is achievable using a mixture of sucrose and 42 DE glucose syrup without sucrose or dextrose crystallisation. However, they point out that there are complications in that whilst sucrose crystallises readily, dextrose monohydrate does not unless seeded. Fondant or confectionery cremes then, normally consist of sucrose crystals within a liquid phase. Obviously the crystal size can affect the eating properties and the largest sugar crystals are normally in the range of 15–45 μ.

So-called dextrose fondants can also be prepared which contain dextrose in the crystalline state. However, dextrose is not the sole ingredient used, as they are often made from dextrose, a high DE glucose syrup and invert sugar. Dextrose fondants are sometimes used to produce peppermints as they produce a noticeable cooling effect in the mouth on dissolution of the dextrose due to the negative heat of solution. The overall moisture content of fondants, typically about 10–12%, is important, because it can obviously have an effect on the percentage of crystalline phase and on the shelf life of the product.

In order to provide a variety of textures it is sometimes desirable to manufacture products with more liquid centres. Although fondants can be produced with jam centres made by double depositing techniques, liquid centres can also be produced by the inclusion of a small quantity of invertase in the centre. On storage this will convert some of the sucrose to invert sugar, i.e. to dextrose and fructose, and lead to a liquid centre by virtue of the greater solubility characteristics of these sugars.

Traditionally fondant cremes are made by depositing the fondant (produced using predominantly sucrose and glucose syrup solids in a ratio of about 4·5:1) into moulds of starch of 6–9% moisture content contained in trays. The starch-filled trays have depressions made in them to produce the moulds into which the fondant is deposited. These are then left to stand at 20–25°C for up to 48 h during which time a moisture transfer from the fondant into the starch occurs leading to fondants of a firmer external texture which can then be shaken from the moulds and enrobed in chocolate. The starch can then be recovered, dried and reused. Following a joint development by Cadbury and Baker Perkins[9] a high proportion of the fondant cremes now produced are deposited directly into metal moulds sprayed with a suitable release agent such as a vegetable oil. This technique is much more efficient and cleaner, although it is important that the crystallisation takes place rapidly after depositing the fondant, and consequently slight changes in recipe and/or processing are required.

Toffees and Caramels

In the manufacture of toffees and caramels use is made not only of the preservative, sweetness, solubility and other attributes of the nutritive sweeteners used in fondants and boiled sweets, but also their ability to be caramelised. The distinction between toffees and caramels is to some extent arbitrary. They are both produced using glucose syrup, sugar, milk solids (which include lactose), fats and sometimes invert sugar, the whole being cooked to yield a product of high total solids content. Differences in the proportions of the various nutritive sweeteners used can influence the shelf life, flavour and texture of the products. The texture is also of course influenced by the residual moisture content which may vary between 6 and 11% w/w. Although residual moisture content is not the only factor which affects the final texture of the product, it nevertheless has a marked influence. High sucrose levels can give rise to a tendency to grain, that is, for crystallisation to occur. Different types of glucose syrup can also influence the texture, low DE glucoses giving a harder product than those of high DE.

As both reducing sugars and casein are found to be necessary for the development of the caramel flavour, Maillard reactions are almost certainly involved. The effects of the presence of butter, which improves the final flavour, are not fully understood. Considerable research has been carried out on the various chemical reactions involved in caramelisation, on the reactions between milk protein and reducing sugars, and on the role of butter, at the Food Research Association at Leatherhead by D. I. Stansell, T. M. Sharp and their co-workers. It is not appropriate in this brief overview of the uses of nutritive sweeteners in confectionery to go into detail but this is not to underestimate the importance of these reactions in developing desirable flavour characteristics, nor the importance of understanding them.

Fudges

Fudges may be regarded as intermediate between toffees and fondants in that, whilst crystallisation in toffees is undesirable, crystallisation is intentional and deliberate in a fudge, the solid phase normally comprising sucrose crystals, fat and milk solids. Lactose crystallisation may occur if significant quantities of lactose are present. Although it does not crystallise readily without being seeded, once initiated it can give rise to large crystals and a coarse texture.

Further Types of Confectionery

There are many other types of confections in addition to those so far considered. These include products such as marshmallows in which a mixture of sucrose, glucose syrup and possibly invert sugar is whipped into a stabilised foam in the presence of a gelling agent; nougat-type products in which the 'base composition' is made by whipping a high boiled syrup including a fat; tabletted products produced by tabletting usually sucrose with a suitable binding agent, or sometimes for mints, dextrose; and panned goods in which other confectionery items or nuts are coated with a hard outer coating through rotation in a pan to which additions of a suitable high sucrose concentrated syrup are made. All of these similarly utilise nutritive carbohydrates in a variety of ways. In the case of jellies and pastilles in which the texture may be achieved by starch or pectin gels, although sucrose and glucose syrup may be included, sucrose may also be used decoratively for dusting.

Detailed information on the manufacture of various kinds of confectionery may be found in textbooks such as those of Minifie[10] and Lees and Jackson.[8]

Other Considerations

The examples considered serve to show that in the manufacture of the great variety of confections that there are, the confectioner has exploited many different properties of the different sugars both singly and in combination. It is interesting to note, however, that when using different nutritive sweeteners (for example, in the manufacture of some fondants or fudges in which controlled crystallisation is important to develop the desired texture) other properties of the same sweeteners may prove to be a disadvantage. Some consumers claim that some fudges and fondants are too sweet. The intensity of different nutritive sweeteners of course varies. Nicol[7] has pointed out the difficulties of conducting sweetness assessments and, quoting from Nieman,[11] has tabulated the relative sweetness at 10% sucrose level of a number of sugars in order of decreasing sweetness as follows: fructose, sucrose and invert sugar, dextrose, maltose and sorbitol, and lactose. The problems involved in the measurement of sweetness have also been considered in detail during a previous symposium in this series.[12]

It is not always possible to interchange different sweeteners so as to reduce the sweetness levels within the confection whilst retaining the desired texture. There would appear to be a need for a product with many of the properties of sucrose but of reduced sweetness.

Reference has already been made to the ability of nutritive carbohydrate sweeteners, through their solubility and resultant increase in osmotic potential and reduction in water activity, to act as preservatives. It is this property, of course, which gives confectionery a long shelf life. Cakebread[13] has summarised in a neat bar chart form the equilibrium relative humidity (ERH) ranges into which different types of confections fall and compared them with the minimum ERH permitting growth of various types of microorganisms. Salmonellae require a high water activity—a vapour pressure above 95%—and bacteria normally require a higher vapour pressure than moulds. Of the various types of confections, fondant cremes are the most vulnerable, particularly to osmophilic yeasts. Great care needs to be taken in formulating them in order to minimise the risks.

CONCLUSIONS

It has been shown that the nutritive sweeteners perform a variety of functions in different types of confections besides the obvious function of providing sweetness. There are limits to which they may be interchanged, in some cases from the practical point of view and the consequences to the characteristics of the product, and in other cases, e.g. in the case of chocolate products within EEC countries, from a legislative standpoint.

REFERENCES

1. COCOA CHOCOLATE AND CONFECTIONERY ALLIANCE (1979). *Confectionery in Perspective*, Cocoa Chocolate and Confectionery Alliance, London, 3–6.
2. *The Cocoa Chocolate and Confectionery Alliance Annual Report 1979–80*, Cocoa Chocolate and Confectionery Alliance, London, 7.
3. *The Food Labelling Regulations 1980*, SI No 1849, HMSO, London.
4. *The Cocoa and Chocolate Products Regulations 1976*, SI No 541, HMSO, London.
5. Council Directive of 24 July 1973 (73/241/EEC) in *Official Journal of the European Communities*, 16 L228, 16 August 23–25 as amended by Council Directives of 1 August 1974 (74/411/EEC); 19 December 1974 (74/644/EEC); 4 March 1975 (75/155/EEC); 20 July 1976 (76/628/EEC); 29 July 1978 (78/609/EEC); 10 October 1978 (78/842/EEC)

and 30 June 1980 (80/608/EEC). Office for Official Publications of the European Communities, Luxembourg.
6. BIRCH, G. G., GREEN, L. F. and COULSON, C. B. (1970), *Glucose Syrups and Related Carbohydrates*, Applied Science Publishers, London.
7. NICOL, W. M. (1979). In: *Sugar: Science and Technology*, Eds. Birch, G. G. and Parker, K. J. Applied Science Publishers, London, 211–30.
8. LEES, R. and JACKSON, B. (1973). In: *Sugar Confectionery and Chocolate Manufacture*, Leonard Hill Books, Aylesbury, Chapter 1, 7.
9. JEFFERY, M. S. (1963). Brit. P. Specification 1050699.
10. MINIFIE, B. W. (1980). *Chocolate, Cocoa and Confectionery: Science and Technology*, Second edn, Avi Publishing Company Inc., Westport, Connecticut.
11. NIEMAN, C. (1958). *Zucher Süsswaren wirtsch*, **11**, 420.
12. SPENCER, H. W. (1971). In: *Sweetness and Sweeteners*, Eds. Birch, G. G., Green, L. F. and Coulson, C. B. Applied Science Publishers, London, 112–29.
13. CAKEBREAD, S. H. (1971). *Manufacturing Confectioner*, **4**, 45–9.

4

Malt and Maltose Syrups

P. D. FULLBROOK

National College of Food Technology,
University of Reading,
Weybridge, Surrey, UK

ABSTRACT

Malt syrups were originally simple concentrates of the water-soluble extract of malt. Malt is traditionally produced by skilfully encouraging and then arresting the natural process of germination. In this two stage process, endogenous enzymes are produced which solubilise the food reserves stored in the grain, and the characteristic flavour components are developed during the subsequent kilning process. The major nutritive components of malt syrups are the disaccharide sugar maltose, various amino acids and peptides, and vitamins. Malt syrups are thus not only a source of carbohydrate sweetness and calorific value, but also (and perhaps more importantly) a source of flavour, natural colour, essential amino acids and vitamins—a truly nutritive sweetener.

Recently the term malt syrup has been (mis)used to imply syrups produced from starch, in which at least one third of the sugar content is maltose. These types of syrups are perhaps more correctly termed maltose syrups.

This paper will review the production of both type of 'malt syrups' in relationship to their applications both as a food ingredient and as a fermentation component for the production of alcoholic beverages (beers and whiskies). Emphasis will be placed on the production of high yield/low cost malt extracts from barley and some of the newer developments in the production of 'tailor made' syrups containing maltose from cereal and tuber starches (i.e. from unmalted materials) which are used in the food and confectionery industries. The significance of cereal and microbial enzymes in these processes will be examined. Finally the economic and legislative background to the production of maltose syrups is considered.

INTRODUCTION

For the purpose of this presentation it is proposed to use the terms 'malt' and 'maltose syrups', in their widest sense to encompass the whole range of extracts and syrups produced from cereals or refined starches, in which maltose is the major carbohydrate component. Extracts produced from whole cereals—and therefore containing nitrogenous components—will be generally referred to as malt or grain syrups, whilst simpler starch hydrolysates will be referred to as starch syrups. Emphasis will be placed on the biotechnological aspects of these developments, principally those of applied enzymology.

Enzyme Systems

Developments in the production of malt and starch syrups over the past 10 years have proved somewhat of a field-day for the applied enzymologist. As a result of fruitful research and development efforts by the enzyme producing companies, currently about a dozen enzymes are commercially available for application in the production of a whole range of industrially important syrups, used mainly by the brewing and food industries. These enzymes are listed[1-15] and classified in Table 1. They can supplement or replace traditional cereal enzymes facilitating a more efficient production of established products (extracts, syrups or nutritional products) or allow their production from alternative, cheaper raw materials. More recently, the development of new enzyme systems and processes[16] has allowed the production of entirely new products for the food and biochemical industries.

Sugars in Nature

Many of the primary nutritive carbohydrates in nature are disaccharides, e.g. maltose, sucrose and lactose. Maltose is produced as a result of germination in cereals or tubers by the action of endogenous β-amylase on the storage carbohydrate, starch, or as a result of digestion in animals (salivary α-amylase). Monosaccharides are also abundant in nature; for example, glucose, fructose and certain sugar alcohols are present in various fruits. Fructose also occurs as a result of insect metabolism (e.g. in honey), whilst glucose is produced directly from the animal storage carbohydrate, glycogen, in the liver and muscles by anaerobic

TABLE 1

CLASSIFICATION OF ENZYMES INVOLVED IN THE (COMMERCIAL) PRODUCTION OF MALT EXTRACT AND STARCH SYRUPS

Type	Enzyme classification Variety	EC number	Common name	Example of production spp.	Substrate specificity	Note
Endo-amylase	thermostable α-amylase	3 2 1 1	bacterial amylase	Bacillus subtilis (amyloliquefaciens)	α-1,4-glucosyl	see Fig. 6
				Bacillus licheniformis	1,4-glucosyl	see Fig. 7
	thermolabile α-amylase:					
	maltogenic		fungal α-amylase	Aspergillus oryzae	α-1,4-glucosyl	see ref. 1
	oligogenic		saccharifying amylase	Bacillus BL 458		see ref. 2
Exo-amylase	glucogenic γ-amylase	3 2 1 3	amyloglucosidase	Aspergillus niger	α-1,4-glucosyl	see ref. 3
				Rhizopus spp.	α-1,6-glucosyl	
	maltogenic β-amylase	3 2 1 2	cereal β-amylase		α-1,4-glucosyl from non R end	
			sweet potato amylase			
			bacterial β-amylase	Bacillus spp.		see refs 4–7
	maltotetrogenic amylase			Pseudomonas stutzeri		
	maltohexogenic amylase			Aerobacter aerogenes		
α-1,6-amylase	pullulanase	3 2 1 41	debranching	Klebsiella aerogenes	α-1,6-maltotriosyl	see ref. 8
	iso-amylase	3 2 1 68	enzymes	Pseudomonas SB 15	α-1,6-heptasacch.	see ref. 9
Endo- + exo-cellulase	β-1,4-cellulase	3 2 1 4	fungal cellulase	Trichoderma reesei	C_x-C_1 = 10	see ref. 10
Endo-glucanase	laminarinase	3 2 1 73	bacterial β-glucanase	Bacillus subtilis	β-1,4-glucosyl	see ref. 11
		3 2 1 39	fungal β-glucanase	Aspergillus niger	β-1,3-glucosyl	see ref. 12
Isomerase	xylulose isomerase	3 5 1 5	glucose isomerase	Bacillus circulans	aldo/keto pentose	see ref. 13
					aldo/keto hexose	
Endo-proteinase	alkaline protease	3 4 21 14	bacterial	Bacillus licheniformis		see ref. 14
	neutral protease	3 4 24 2	protease	Bacillus subtilis		see ref. 15

metabolism, as a response for instant energy. Trisaccharides arise as a result of the degradative action of microorganisms on structural carbohydrates, e.g. cellotriose from cellulose and maltotriose from α-1,6-carbohydrate polymers such as pullulan.

MALT SYRUPS

Composition

Historically, maltose-containing extracts have been prepared and used for thousands of years in the worts used by brewers and distillers in the production of their respective beverages and in the doughs used by housewives and bakers in the preparation of bread. Although maltose is the most important component in malt extracts, the other components also have significant nutritive value. Table 2 indicates the composition of the major fractions of carbohy-

TABLE 2
COMPOSITION OF THE MAJOR FRACTIONS OF WORT CARBOHYDRATES

Grist component	Solids (%)	Grist (%)	
		Mash A	Mash B
Malt type 1	100	100	—
Malt type 2	96	—	79·4
Maize grits	90	—	14·9
Glucose	87	—	5·7

Wort component	DP	Total wort carbohydrate (%)	Wort (g/100 ml)	Solids (%)	Fermentable sugar (%)
Dextrins + pentosans	5	22·2	—	—	—
Maltotetrose	4	6·1	—	—	—
Maltotriose	3	14·0	1·56	13·9	21
Maltose	2	41·1	4·51	40·2	61
Sucrose	2	5·5	—	—	—
Glucose	1		1·31	11·7	18
Fructose	1	8·9	—	—	—
Total		97·8	—	—	—
Fermentable sugar		69·5	7·38	65·8	100
Extract		103 lb/Qr		10·75°Balling	

Data from References 17 and 18.

drates present in malt syrups (derived as brewers' wort)[17,18] and shows maltose as the major carbohydrate constituent. This is particularly significant if the extract is to be fermented, since high concentrations of glucose tend to inhibit fermentation to ethanol.[19] Table 3 gives general data on the other important fraction of malt extract—the nitrogenous components.[20,21] Of these the vitamins[22–30] are important, acting as they do as co-factors for enzymes involved in body metabolism. The amino acid composition[18,31] of some representative malt extracts are given in

TABLE 3
NITROGENOUS COMPONENTS OF WORT/MALT EXTRACT—GENERAL DATA

Component	
Nitrogenous materials equivalent to	5–6% wort solids
	30–40% total nitrogen in malt
Permanently soluble fraction	94% total soluble wort nitrogen
Amino acids	40% permanently soluble nitrogen

Other total soluble nitrogen fractions include:

Basic materials	choline
	betaine
	ammonia
Amino acid polymers	peptides
	undegraded proteins
Nucleoprotein components	purine + pyrimidine bases
	nucleosides
	desoxynucleosides, from barley DNA

Vitamin components comprise:

Vitamin		Concentration		Reference
		$\mu g/g$ dry wt malt extract	$\mu g/100$ ml wort $(sg = 1040)$	
(meso-Inositol)		1890	18.9×10^3	22
Thiamin	B_1		60	23
Riboflavin	B_2		33–46	24, 25
Folic acid		0·1–1·0	—	26
Nicotinic acid			1000–1200	27
Pantothenic acid			45–65	28
Pyridoxamine	B_6		85	29
Biotin			0·65	30

Data from References 20 and 21.

Table 4. It can be seen from the table that the malt extracts contain significant amounts of the so-called essential amino acids. Of the amino acids listed, malty flavours are associated with the amino acids valine and leucine,[32] although others, particularly proline,

TABLE 4
AMINO ACID COMPOSITION OF WORT/MALT EXTRACT

Grist component	Solids (%)	Grist (%)		
		Mash A	Mash B	Mash C
Malt type 1		100	—	—
Malt type 2	96	—	79·4	51·0
Barley	85	—	—	28·2
Maize grits	90	—	14·9	15·0
Glucose	87	—	5·7	5·8
Microbial enzyme		—	—	+
Mash type		Infusion, 65°C	Decoction (see Fig. 1)	
Mash time (min)		180	200	200
Amino acid		(μg amino-N/g dry malt)	$(mg/l)^a$	$(mg/l)^a$
Total		1470·7	1049·5	861·0
Aspartic acid		43·0	41·3	38·4
Threonine+asparagine[b]		200·9	89·8	85·1
Serine			42·4	31·7
Glutamic acid		53·6	35·0	45·0
Proline		455·6	205·0	151·1
Glycine		28·3	24·9	18·7
Alanine		62·3	61·2	48·9
Valine[b]		99·2	58·0	43·3
Methionine[b]		19·7	15·5	12·8
Iso-leucine[b]		49·1	37·7	28·6
Leucine[b]		101·4	70·0	60·2
Tyrosine		52·4	53·2	43·2
Phenylalanine[b]		76·8	66·0	54·5
Lysine[b]		80·3	54·4	45·6
Histidine[b]		38·1	32·6	22·2
Tryptophan[b]		29·5	33·4	31·1
Argenine		80·5	91·2	67·9
γ-Aminobutyric acid			37·8	32·6

[a] Extract = 10·7°Balling.
[b] Essential amino acids.
Data from References 31 and 18.

can potentiate these malty flavours. Aliphatic amino acids such as glycine and valine tend to produce aromas characteristic of bread and are useful constituents if the malt extract is to be used as an ingredient in the baking of bread. These flavours arise, together with colour components, by interaction of the amino acids with sugars during the production of malt or malt extract.[33]

Traditional Production of Malt Syrups

Malt syrups were originally prepared as simple concentrates of the water-soluble extract of malt. Malt is traditionally produced by first initiating the germination of grain—usually barley.[34] The rootlets are then trimmed from the sprouting barley and the grain skilfully heat-treated in a subsequent kilning process. Basically the initial effect of the germination process is to mobilise the stored carbohydrate reserves by encouraging enzyme development. The kilning process stops further hydrolysis by evaporation of water, whilst leaving most of the newly developed enzymes potentially active. In this partially dehydrated form, malt is microbiologically stable. Also, during kilning, certain flavour components are developed as indicated above.[32] Removal of the rootlets avoids the extraction of bitter nucleic acids during processing.

The extract is produced by assisting the further hydrolysis of the grain arrested during the kilning process. Malt is incubated in aqueous suspension to form an aqueous extract, which is separated from the insoluble residue by filtration. The process is referred to as mashing of the grains (Fig. 1). Fermentation of this syrupy extract will lead to the production of beer, whisky or vinegar depending on subsequent operations, whilst stabilisation or simple concentration will produce malt extract. This is sold for various applications (Table 5). Further processing with other ingredients creates food products—typically, a base for health drinks. Representative analyses of such products are given in Table 6.

Biochemistry of Malting

The biochemical changes involved in the production of malt and malt extract (wort/syrup) from barley can be deduced from Fig. 2.[15] This represents a semi-quantitative schematic comparison of the relevant components of the materials. Malting is essentially an enzyme producing process, the result of which is that a proportion of the starch is hydrolysed to dextrins and most of the high

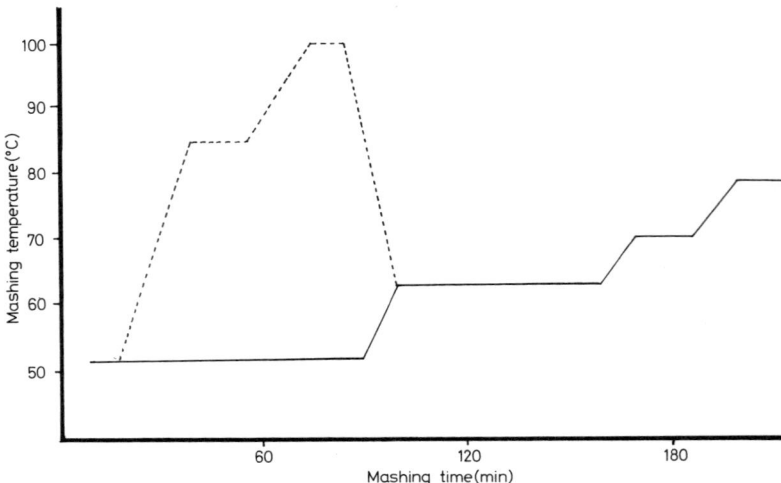

FIG. 1. Mash programme used for production of wort and malt extract, - - - Decoction vessel; ——— mash tun. Taken from Reference 18.

TABLE 5
SOME USES OF MALT EXTRACT/SYRUPS

Function	Speciality	Application
Enzyme source	β-amylase	Production of maltose syrups Starch hydrolysis applications Flour improvement—baking industry
Vitamin source	B complex	Nutritional supplement
Flavour Food ingredient	'Malty tastes' Bulking agent	Food products/health drinks
Fermentations	Barley/malt syrups	Brewers' sugars Home brewing kits

molecular weight β-glucan carbohydrate component is hydrolysed to low viscosity dextrins. Formation of amino acids from insoluble protein also occurs, but during the kilning process, part of the peptidase and β-glucanase activities will have been destroyed.[35] The surviving enzyme activities are effective during the mashing process, completing the solubilisation of carbohydrate and protein components to produce wort extract. These enzymes are com-

TABLE 6
COMPOSITION[a] OF SOME COMMERCIAL MALT/ADJUNCT BASED FOOD PRODUCTS

Ingredients	Health drinks				Breakfast cereals		
	Horlicks (Beecham)	Malted drink (Waitrose)	Ovaltine (Wander)	Bournville (Cadbury)	Experimental preparation (%)	Corn flakes (Kelloggs)	Special 'K' (Kelloggs)
Malt extract	x	x	x	x	8	x	x
Barley extract		x	x				
Wheat flour	x				48·4[b]		
Wheat gluten							x
Wheat germ (defatted)							x
Rice							x
Maize						x	
Full cream milk			x				
Skimmed milk	x	x	x	x	22·8		x
Whey powder	x	x			8·1		
Egg			x	x			
Vegetable fat	x	x			8·5		
Fat reduced cocoa			x	x			
Sugar	x	x	x	x	6·5	x	x
Glucose syrup solids			x	x			
Lecithin		x					
Flavouring				x			
Salts	x	x	x	x	1·4	x	x

Analysis (units/100g)

Protein, N × 6·25(g)						7·5	19·0
Fat(g)							1·0
Available CHO(g)							72·7
Energy(kJ)						1480	1500
Vitamin A(μg)	0·44						
B1(mg)	0·85		x			1·0	1·2
B2(mg)	1·06					1·5	1·7
B6(mg)						1·8	2·2
Niacin(mg)	11·27					16·0	18·3
D/D3(μg)	1·55		x			2·8	2·8

[a] Manufacturers' data.
[b] Wheat extract at 10% DS.
x = ingredient present.

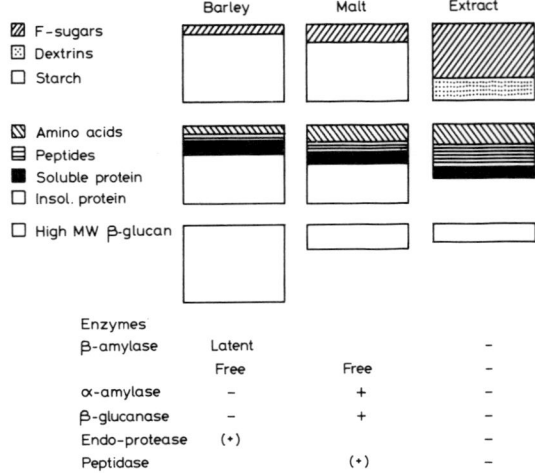

FIG. 2. Comparison of barley, malt and malt extract. The transformation of starch, protein, glucan and enzymes. Data from Reference 15.

pletely destroyed during the evaporation stages of the malt extract production.

Malting Problems

As well as losing enzyme activity during the production of malt, a proportion of the extractable material is also lost in the germinating rootlets causing a loss of 6–10% of the dry matter.[36] Coupled with the increasing energy and storage costs in modern malt production, this makes malt a relatively expensive commodity, usually around 50% more expensive on a direct weight basis than barley (~30% more expensive on an equal extract basis). In order to reduce raw material costs, particularly when using premium quality highly modified malts (high amylolytic activity), a proportion of the malt is usually replaced by a cheaper non-malted adjunct—traditionally a low-nitrogen starch source such as maize grits or rice, or more recently by unmalted grains or flours such as barley or wheat. Conversely, if the malt is of poor quality and low diastatic power, it will be necessary to supplement the enzymatic activity of the malt with added enzymes in order to obtain sufficient extract. Following the dry summer of 1976 in the UK, malt was either generally of poor quality, or scarce and particularly expensive, so that use of microbial enzymes to improve the quality of the malt during the mashing process found wide application.

Figure 3 shows the effect of fungal and bacterial β-glucanases on a laboratory scale, using synthetic 'poor quality' wort produced by extracting milled barley with thermostable endo-amylase.[11] Glucanase enzymes were incubated with wort for 30 min at the temperatures indicated, before being cooled to 25°C for viscosity measurements. Results indicated a better performance by the bacterial enzymes as compared to the fungal enzymes. This was confirmed in a full scale production trial (Table 7) in which the added bacterial enzyme not only considerably reduced filtration time, but also increased yield by 2%. Table 7 thus illustrates the advantage in using microbial enzymes to improve malt quality. An additional problem is that unless the kilning process is carefully controlled, production of nitrosamines can occur. These have been shown to have carcinogenic properties and are a subject of current concern. Use of barley as the source of extract overcomes this problem.

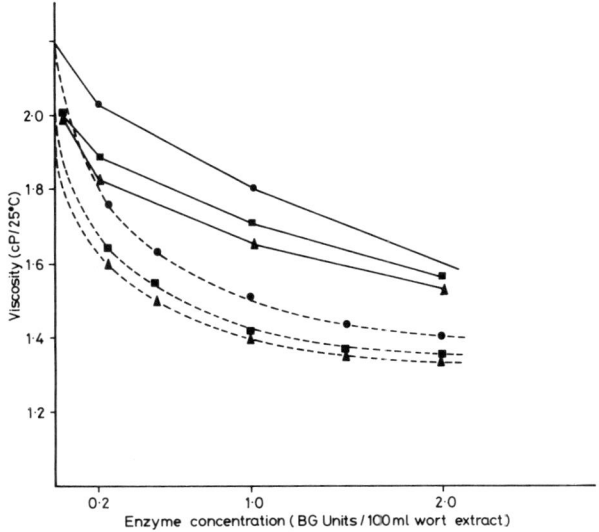

FIG. 3. Effect of microbial β-glucanase enzymes on poor quality grain extract. Fungal enzyme, pH 5·5/30 min, extract = 11·4° Balling:—●—70°C;—■—50°C;—▲—60°C. Bacterial enzyme, pH 5·7/30 min, extract = 9·9° Balling:--●--70°C;--■--50°C; ---▲---60°C. Data from Reference 11.

TABLE 7
100 hl SCALE PRODUCTION TRIAL FOR MALT IMPROVEMENT

Parameter	Control	Trials
Poor quality malt(% grist)	75	75
Maize adjunct(% grist)	25	25
Bacterial β-glucanase(% malt)	0	0·025
Number of runs	6	6
Wort concentration(°Balling)	10·8	10·8
Wort volume (hl)	99·4	101·5
Filtration time(h)	3–3·5	~2

Data from Reference 18.

Alternative Production of Malt Extract

As malt is essentially an intermediate in the production of malt extract, one can extrapolate back from high diastatic malts, through a range of increasingly poorer (cheaper) grades of malt, eventually to unmalted barley as the source of extract. In order to maintain wort quality in this transition, it is necessary to increase proportionately the quantity and type of microbial enzymes added to the mash. Interest in this possibility of 'barley brewing' developed during the 1960s as is indicated by the extensive literature of this period.[37-51] The main process changes necessary to exploit this development are at the milling stage, since barley is mechanically more difficult to crush and mill than malt. This difficulty is overcome by adopting cold-steeping and wet-milling of the barley. Investment costs in new milling facilities are soon recovered because of the lower costs of raw materials and the better control achieved over the mashing process. The potential economic advantages attained by replacing a proportion of the malt by unmalted grains or adjuncts supplemented by microbial enzymes are illustrated in Tables 8 and 9. Using the values given for a brewery in Belgium (July 1979) Table 8 shows that a 50% replacement of malt with barley supplemented by microbial enzymes would give a cost saving of around 15·5 BFr/hl wort extract, or reduce the raw material costs by about 10%. Table 9 indicates approximate cost savings which could have been realised for a UK brewery in 1980. It has been assumed that an extract of 10°P* would be produced,

*Plato, a scale based on the specific gravity at 20°C.

TABLE 8
ECONOMIC ADVANTAGE OF MALT REPLACEMENT

Cost saving per hl extract in replacing a proportion of the malt with unmalted cereals supplemented with microbial enzymes can be estimated using the following expression:

$$\text{Saving} = p\left\{y\left[\frac{M}{m} - \frac{B}{b}\right] - \left[\frac{3Cy}{1000b} + \frac{T(100-x)}{1000a}\right]\right\} \text{ Currency units/hl}$$

Symbol	Example[a]
p = required extract in wort (°P)	10·5
y = % extract derived from barley (15–67)	
M = Malt cost/kg	12·0 BFr
m = % extract derived from malt	80
B = cost of barley/kg	7
b = % extract derived from barley	65
C = cost of mash tun enzymes	
T = cost of decoction vessel enzymes	
a = % extract yield from starch in adjunct	82
x = % extract derived from malt	85
Hence cost saving = $0.37y - 0.25$ BFr/hl	

[a] Values given for a brewery in Belgium (July 1979).

with 10% of the total grist extract being derived from glucose syrup (copper adjunct) of 97% extract yield. Malt I is assumed to be a highly modified malt of 81% extract yield commanding a premium price, whilst malt II is of medium quality with 77% extract yield costing 18% less. Replacing up to 80% of the malt by barley could have led to a cost saving of £337/10^3 hl of wort, or an annual saving of around one third of a million pounds/annum for a brewery of 10^6 hl capacity. Increased plant capacity can also be achieved using microbial enzyme supplements to shorten mash cycling times, usually at the filtration stage. This allows an extra mash cycle to be made every 24 h, as illustrated in Table 10. A similar capacity increase was achieved by a UK brewery during 1977.

Modern Malt Extract Production

Several brewing groups in the UK have taken advantage of these developments over the past years, some deriving up to 80% of their extract from replacing malt with unmalted adjunct, although the response of brewers and distillers in other countries has generally

TABLE 9

POSSIBLE COST SAVINGS IN PRODUCTION OF MALT EXTRACT BY SUBSTITUTION OF A PROPORTION OF MALT WITH ADJUNCT

Malt Replacement (%)	Grist composition Percentage					Actual weight (tonnes per 1000 hl wort)				Total raw material costs (excluding sugar) (£)			Cost saving	
	Malt I	Malt II	Barley	Maize		Malt I	Malt II	Barley	Maize	Grist	Enzyme	Total	(£/tonne grist)	(£/1000 hl wort)
0	80	—	0	10		9·88	—	—	1·2	1972	0	1972	0	0
33	53·3	—	26·7	10		6·58	—	4·11	1·2	1805	32	1837	11·35	135
50	40	—	40	10		4·94	—	6·15	1·2	1703	48	1751	17·98	221
75	20	—	60	10		2·47	—	9·23	1·2	1596	73	1669	23·49	303
80	16	—	64	10		1·98	—	9·85	1·2	1571	78	1649	24·79	323
0	40	40	0	10		4·94	5·20	—	1·2	1895	5·46	1901	6·26	71
33	26·7	26·7	26·7	10		3·29	3·46	4·10	1·2	1754	35·63	1790	15·10	182
50	20	20	40	10		2·47	2·60	6·15	1·2	1683	50·73	1734	19·16	238
75	10	10	60	10		1·24	1·30	9·23	1·2	1577	74·37	1651	24·75	321
80	8	8	64	10		0·99	1·04	9·85	1·2	1556	79·09	1635	25·77	337

TABLE 10
EXAMPLE OF CAPACITY INCREASE USING ENZYMES

Brewhouse capacity nominally	1000 hl/day
Normal mash cycle	4 h (limited by filtration stage)
Usual production capacity	800 hl/day
Required output increase	$\frac{1}{3}$

Mash cycle of 4 h allows 6 cycles/24h
Normal output per cycle   = 800/6   = 133·3 hl
Maximum output per cycle = 1000/6   = 166·7 hl
Required output per cycle $(800 + 800/3)/6 = 177·8$ hl

Experiments showed that use of β-glucanase enzymes reduced filtration time to just over 3 h 15 min, allowing an extra mash cycle/24 h

Target output $= 177·8 \times 6 = 1067$ hl
Actual output $= 166·7 \times 7 = 1167$ hl $= 9\%$ above requirement
Maximum allowable filtration time $= 24/7 = 3$ h 26 min

	Capacity (hl/day)	
	800	1167
	Grist cost/day (£)	
Malt	1260	1840
Adjunct	416	606
Enzyme	0	10
Total	1676	2456
Cost/hl extract (£)	2·095	2·105

Increase in raw materials $= \frac{1}{2}\%$. Increase in capacity $= 46\%$.

been more enthusiastic.[52] This is probably due to the poorer qualities of malt available to them, or to the fact that there is less consumer pressure for 'traditional' products. In contrast, however, most of the malt extract now produced worldwide is derived by processing barley in 30–80% replacement of malt. The enzymes used in this are those previously listed in Table 1. Thermostable endo-amylase is used to liquefy the starch in the grain, enabling the native cereal β-amylase to hydrolyse a proportion of the soluble dextrins to sugars. Bacterial β-glucanase hydrolyses the β-1, 4:β-1, 3 carbohydrates to low viscosity dextrins, allowing a more efficient separation of the extract. It also lowers the energy costs in evaporating reduced volumes of sparge water, and promotes easier evaporation of water from the extract.

Bacterial neutral proteinase solubilises proteins and partially hydrolyses them to produce flavour,[32] colour[33] and a satisfactory amino acid spectrum.[18, 31] Another important function of neutral proteinase is to activate cereal β-amylase, part of which is present in ungerminated barley as an inactive pro-enzyme.[53, 54] An example of the use of microbial enzymes in producing malt extract will now be described.

Laboratory trials indicated that it was possible to produce a grain syrup from 100% barley using microbial enzymes as mash supplements. On the basis of these experiments, a full-scale set of trials using 50% malt replacement were conducted in a Danish brewery equipped with a Steinecker brewhouse plant of 100 hl capacity, incorporating wet-milling. Barley and malt were mixed and steeped in water for 15 min at 45–50°C and then milled. The mash programme used was the one indicated in Fig. 1. Enzymes were added to the mash at 50°C prior to the commencement of the mash cycle. Although maize grits were liquefied with a proportion of the malt in the control syrup production—as is traditional in the production of brewers wort—this was substituted with bacterial α-amylase in the trial runs. The malt substituted in this manner was added to the main mash. The wort at mash-off was filtered in a traditional lauter tun before boiling and concentrated to 80% solids under vacuum evaporation. The results from the test laboratory syrup preparation and the full scale control and trial syrup runs are summarised in Table 11. The results indicate that a malt extract of similar composition to the control (malt/maize, syrup 2) could be produced when replacing 50% of the malt with barley, supplemented by microbial enzymes. Fermentable extract and nitrogen levels were very similar, and filtration times better, than in the control mash. The trials also showed convincingly that it was possible to replace the malt used for liquefaction of the maize adjunct by α-amylase to give better use of the malt. The use of amyloglucosidase (syrup 4) increased glucose levels at the expense of maltose and maltotriose levels. Higher nitrogen levels in this syrup may be a consequence of a contaminating proteinase in the enzyme preparation used. The α-amylase used also contained a high proportion of endo-β-glucanase (although dosed on amylase activity). The n-proteinase had a contaminating alkaline proteinase activity of no significance in the mash.

TABLE 11
LABORATORY AND PLANT SCALE PRODUCTION OF MALT SYRUPS

Raw materials	Laboratory test		Full scale plant trial (100 hl)					
Grist	syrup 1 (kg)	(%)	syrup 2 (kg)	(%)	syrup 3 (kg)	(%)	syrup 4 (kg)	(%)
Malt for main mash	—	—	1060	70·2	575	39·8	575	39·8
Malt for maize liquefaction	—	—	70	4·6	—	—	—	—
Barley (9·1% protein)	0·05	100	—	—	490	33·9	490	33·9
Maize grits (total solids)	—	—	380	25·2	380	26·3	380	26·3
Enzyme supplement								
α-amylase (liquefaction)						0·4		0·4
α-amylase (main mash)						0·86		0·86
γ-amylase (amyloglucosidase)						—		1·00
Proteinase						0·08		0·08

Composition of syrup produced (expressed as % of total solids)

(Total solids)	79·5		80·0		80·0		80·0	
Glucose	6·4		6·38		4·73		6·98	
Maltose	52·0		39·9		44·0		43·1	
Maltotriose	—		16·1		14·7		12·5	
Σ Fermentable sugar	—		62·4		63·4		62·6	
Total N	0·8		0·47		0·44		0·48	
α-amino N	—		0·12		0·09		0·11	
α-amino N total N	—		24·4		20·6		21·7	
Formol N	0·2							
Filtration time (min)	—		150		120		120	

Data from Reference 18.

Process Improvement

More recent experiments on the production of malt syrups and extracts have concentrated on the use of β-glucanase and cellulase enzymes to improve extract yield, decrease run-off times and lower condensing times—i.e. improve process efficiency and lower processing costs further. In one series of trials, laboratory investigations had indicated beneficial effects when using cellulase enzymes and full scale trials were designed and run to investigate improved process economy in an already established enzyme-assisted system. The plant processed 200 tonnes cereals per day in a 1200 hl plant using fast mashing and filtration cycles, to produce eventually a high gravity extract. Relevant process details and results of trials are shown in Table 12.

TABLE 12
1200 hl SCALE PRODUCTION TRIAL FOR ADJUNCT/MALT EXTRACT

Cereal grist composition		Percentage of total		Enzyme (% w/w on main adjunct)
component	kg	grist	extract	
Malt	7300	29·5	32·5	—
Barley	8534	34·5	32·0	—
Wheat	2100	8·5	9·7	—
Enzyme, control	52	—	—	0·61
Enzyme, test	32	—	—	0·37
Grist:liquor ratio	1:6·7 litre mash			

Enzyme dosing activities		Enzyme units × 10³/mash			
	control	trials: 1.	2.	3.	4.
α-amylase(bact. KN)[a]	1320	2990	2990	2990	2990
β-glucanase(fung. BG)[a]	330	193	193	—	193
β-glucanase(bact. BG)[a]	210	8050	8050	8050	8050
endo-cellulase(C_1)[a]	70	450	450	—	450
exo-cellulase(C_x)[a]	968	3533	3533	—	3533
endo-proteinase (n-Anson)[a]	4·34	4·34	4·34	4·34	4·34
Results					
No. of runs	∞	1	5	6	11+
Extract(°Balling)	15·5	satisfactory	—	—	improved
Run off(min)	90	88	89	118	89
Economy estimate (% saving)	—	—	?	NV	11·5

[a] Enzyme units: Standard Novo units, as defined in specifications.
NV: Not valid.

The results verified that cellulase was necessary to maintain fast run-off. The trial enzyme runs also tended to improve the yield and offered the possibility of at least a 10% enzyme cost saving over the control runs. The relatively higher enzyme activity amounts used in the trial runs (compared to the controls) gave further indication that economy could be improved by reducing incrementally enzyme levels in additional trials.

A second series of trials was designed for a similar but smaller plant processing 45 tonnes cereals per day. The mashing/filtration unit was coupled with a multiple-effect vacuum evaporator to produce an 80% DS malt extract, which was processed further to make a proprietary food product. Unlike the previous plant, a high grist/liquor ratio was used to minimise evaporation costs, and hence a more conventional mash cycling programme operated. In order to maximise output, run-off and condensing times are critical. The object of the trials was thus to maintain these factors, whilst improving overall economy by lower enzyme costs, improved extracts or, preferably, both. Laboratory trials using factory materials had indicated a 5% improvement in yield by using an enzyme combination designed for the trial. The results of the trial are summarised in Table 13.

Data showed that trial run 2, with elevated enzyme levels, gave an improved performance over the control enzyme system. The enzyme combination used in trial 3 offered no significant advantage and indeed was more expensive in enzyme costs. Detailed wort analyses of this run are given in Table 14 and are compared with pre-trial laboratory runs, indicating the value of preliminary investigations and enzyme design work.

MALTOSE SYRUPS

As indicated in Table 5, malt extract can be used as an enzyme source for the production of maltose syrups from starch. Interest in maltose syrups produced by these means is centred not on their sweetening effects (as is the case for invert or fructose syrups) but on a whole range of structural and physiological properties,[55] such as high fermentability, osmotic and stabilising effects and physical behaviour when used as ingredients in confectionery and other manufactured food products. They are valuable products as they

TABLE 13
250 hl SCALE PRODUCTION TRIAL FOR HIGH EXTRACT MALT SYRUP

Cereal grist composition component	kg	Percentage of total grist	extract	Enzyme (%w/w on adjunct)
Malt	2625	47·7	52	—
Barley	2875	52·3	48	—
Enzyme, control	15	—	—	0·52
Enzyme, test	9·9–10·8	—	—	0·35
Grist:liquor ratio	1:3·5 litre mash			

Enzyme dosing activities	control	Enzyme units × 10^3/mash trials:1.	2.	3.
α-amylase (fung. KN)[a]	503	403	726	514
α-amylase (bact. KN)[a]	568	454	755	534
β-glucanase (fung. BG)[a]	125	104	179	500
β-glucanase (bact. BG)[a]	55	45	2037	1440
endo-cellulase (C_1)[a]	4	3	424	196
exo-cellulase (C_x)[a]	303	242	3327	1535
Alkaline protease(Anson)[a]	0·82	0·84	0·10	1·22
Neutral protease(Anson)[a]	0·11	0·12	2·12	0·89
Results				
Mash volume (hl)	192·5 ± 5%	—	190	195
Extract (°Balling)	19	—	20·5	20
Run off (min)	180	—	<180	<180
Condensation (min)	180	—	150	180

[a] Enzyme units: Standard Novo units, as defined in specifications.

TABLE 14
MALT EXTRACT WORT ANALYSIS

Parameter	Laboratory test run	Factory trial	
Extract (°Balling)	20·8	19·6	
R-sugar (= g glucose/100 ml)	8·2	7·1	
Glucose (% of total)		6·89	8·9[a]
Maltose (% of total)		50·56	65·6[a]
DP_3 (% of total)		19·64	25·5[a]
DP_4 (% of total)		22·91	
Viscosity (cP (10° Brix/25°C))	1·38	1·40	
α-amino N (mg/100 ml)	52	55	

[a] Percentage F-sugar.

do not exhibit a tendency to crystallise (as do glucose syrups) and are relatively non-hygroscopic. They are currently finding a use in frozen dessert formulations where they control crystal formation, as well as in the baking and brewing industries where their high fermentability is the important property.

Traditional Maltose Syrup Production

The original maltose syrups were produced by saccharification of acid-liquefied (thinned) starch with malt extract (β-amylase). Acid-hydrolysis of starch proceeds by a random chemical attack on the starch molecules and oligosaccharides are formed leading to a regular distribution of concentrations of oligomers in order of degree of hydrolysis. This is illustrated by the gel chromatogram of a 34 DE acid-hydrolysed corn starch syrup shown in Fig. 4(a). As the hydrolysis proceeds the dextrose equivalent increases. (Syrups

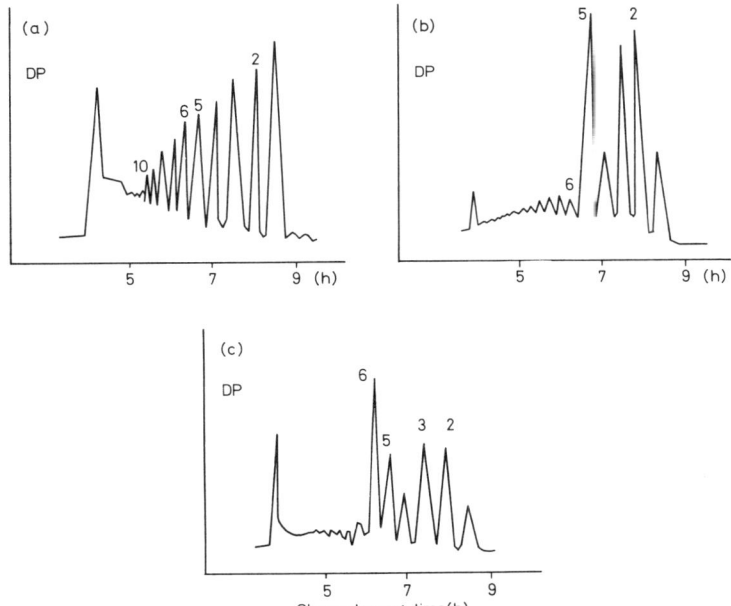

FIG. 4. Gel chromatograms of hydrolysed starch solutions. (a) 34DE acid hydrolysis; (b) hydrolysis using *B. licheniformis* α-amylase for 24 h; (c) hydrolysis using *B. subtilis* α-amylase for 24 h. Data from Reference 2.

produced by acid-hydrolysis are traditionally characterised by their dextrose equivalent, i.e. their equivalent concentration of reducing sugar expressed in terms of glucose.) The degree of hydrolysis effected by acid is controlled by adjustment of the reaction time under pressure and at elevated temperature (120–140°C) and the aim is to produce a product of median DE value, e.g. 15–18. Products of too low DE value are liable to retrogradation— repolymerisation of partially degraded starch to an intractable, stable by-product. Hence these syrups will contain a proportion of glucose, which reduces their value as a starting point for production of maltose syrup, since this automatically reduces the amount of potential maltose which could be formed by the action of β-amylase.

Low-glucose-containing Maltose Syrups

Development of thermostable bacterial endo-α-amylase offered an alternative to acid liquefaction of starch.[56,57] Starch is first gelatinised by heat treatment (usually in the range 50–70°C). This disrupts the areas of macromolecular crystallinity and induces aqueous solubility, facilitating the attack of amylolytic enzymes. Contrary to previously held views, this step is not obligatory for enzymatic hydrolysis which can, like phosphorolysis of starch, proceed at sub-gelatinisation temperatures, although requiring higher enzyme dosages and much longer reaction times.[58] The specificity of endo-amylolytic enzymes causes production of a series of oligosaccharides whose average degree of polymerisation (number of glucose units) is approximately 10.

Hydrolysates produced by amylases, obtained by the use of different types of thermostable endo-amylases (produced by different Bacillus species), have different carbohydrate spectra[2] as illustrated in Fig. 4(b) and (c). The enzyme from *Bacillus licheniformis* produces initially maltohexose which is subsequently hydrolysed, giving mainly maltose, maltotriose and maltopentose, whilst the enzyme from *Bacillus subtilis* is unable to hydrolyse maltohexose further. Hence these enzymes, particularly the *B. subtilis* enzyme, offer potentially a better substrate for saccharification to maltose than do the acid-hydrolysed starches.

Improved Maltose Syrups

A second enzyme development of significance in the production of maltose-containing syrups, has been the discovery that thermolabile (fungal) endo-amylase has a much broader substrate specificity than the corresponding bacterial enzymes just discussed. Thus if this enzyme is left in contact with a suitable starch or starch hydrolysate of DE 10–20 for a long enough period, it will hydrolyse α-1,4-oligosaccharides right down to maltose and maltotriose.[59] This is illustrated in Fig. 5. This enzyme is therefore a cheap alternative to the cereal β-amylase (malt extracts) previously used.

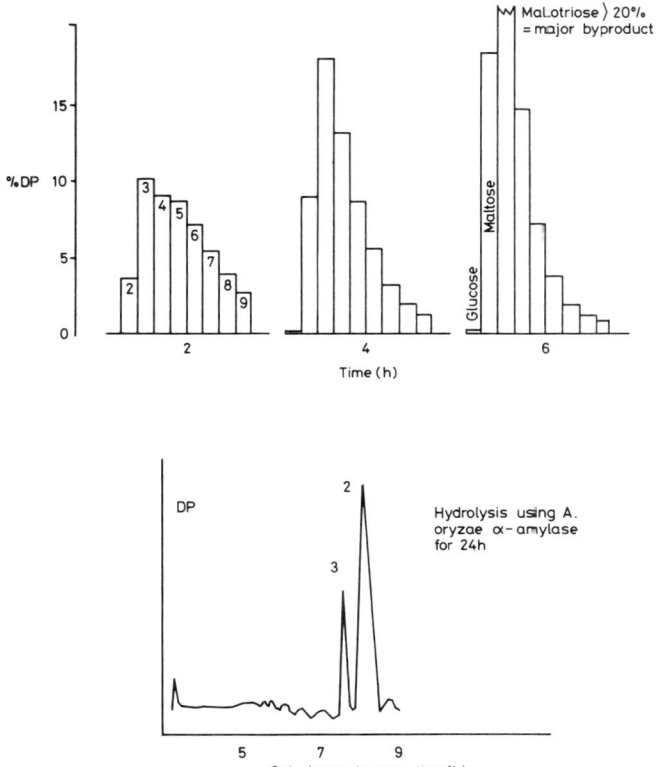

FIG. 5. Hydrolysis of starch by maltogenic endo-amylase revealed by gel chromatography of reaction samples taken out of a reaction mixture of 22% starch with Fungamyl® 800 L at pH 5·5 and at 55°C. Concentration of maltose after 24 h = 22%. DP = degree of polymerisation. Data from References 57 and 2.

Applying these two developments, it is possible to produce maltose syrups containing a relatively high proportion of maltose with a low content of glucose, thus improving on their levels in malt syrups (Tables 2 and 15). The use of two microbial endo-α-amylases from *Bacillus subtilis* and *Aspergillus oryzae* to produce such syrups is illustrated in Fig. 6.[60]

TABLE 15
IMPROVED MALTOSE SYRUP PRODUCTION USING ALL ENZYME PROCESS

Substrate DE	Liquefaction agent	Saccharification enzyme dosage	DE	Product analysis glucose (%)	maltose (%)	triose (%)
38	Acid	75	54	18	41	24
38	Acid	150	54	19	44	19
20	Acid	150	51	10	54	22
20	Enzyme	150	48	4	59	24

Reaction conditions
Substrate 45% w/w hydrolysed corn starch
Saccharification: temperature, 55°C
　　　　　　　　　pH, 5·0
　　　　　　　　　time, 42 h
　　　　　　　　　enzyme, thermolabile endo-amylase *A. oryzae* amylase, dosed at indicated g/tonnes starch

Data from Reference 60.

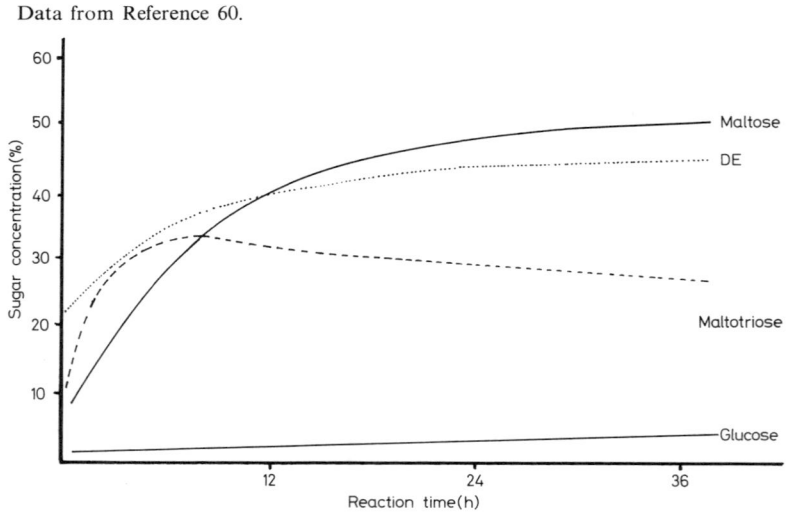

FIG. 6. Change in sugar spectrum during saccharification with fungal α-amylase → maltose syrup. Substrate: 20 DE enzyme liquefied corn starch at 40% DS. Reaction: 50°C; pH, 5·0; enzyme = *A. oryzae* endo-amylase dosed at $1·6 \times 10^5$ units (100 g)/tonnes starch. Data from Reference 60.

High Conversion Syrups

An enzyme previously mentioned in connection with early experiments to produce malt extract, and one which is used extensively in the production of glucose and fructose syrups, is amyloglucosidase. This is extracted from *Aspergillus niger* In contrast to the *Aspergillus oryzae* enzyme, amyloglucosidase is an exo-α-amylase and will produce glucose from oligosaccharides. As both these enzymes have similar pH optima (Fig. 7) they can work simultaneously on partially hydrolysed starch substrates to pro-

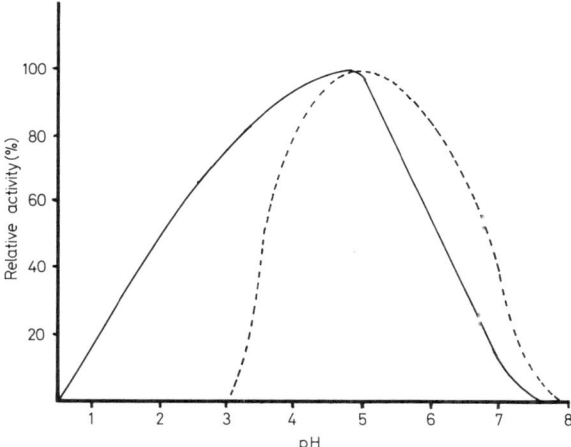

FIG. 7. pH activity curves for saccharification enzymes. ——— Glucogenic exo-amylase from *A. niger*; ---- thermolabile endo-amylase from *A. oryzae*. Data taken from manufacturers' information literature.

duce high-conversion syrups. These syrups have the highest possible DE but are stable enough to resist crystallisation at 4°C and 80–83% DS. They are widely used in the soft-candy, baking, soft drinks and canning (food preservation) industries and because of their low hygroscopicity and high fermentability also find use in the brewing industry. A typical specification is shown in Table 16 and examples of the production of such syrups are given in Figs 8 and 9.[60]

High Maltose Syrups

The following paper describes how the efficient depolymerisation of starch to its constituent monomer, glucose, is effected on a

TABLE 16
SPECIFICATIONS OF ENZYME-PRODUCED SYRUPS

	Glucose	Maltose	Maltose	High maltose	High conversion
Liquefaction	*B. licheniformis* α-amylase	*B. subtilis* α-amylase	*B. licheniformis* α-amylase	*B. licheniformis* α-amylase	acid
De-branching	*A. niger* amyloglucosidase	—	—	*K. aerogenes* pullulanase	acid
Saccharification	*A. niger* amyloglucosidase	*A. oryzae* α-amylase	barley β-amylase	barley β-amylase	*A. niger* + *A. oryzae* enzymes
DE	96–98	40–50	—	—	63–67
Glucose (%DS)	95–97	2–7	0·7	2·0	34–40
Maltose (%DS)	1·5–2·0	45–60	59·4	66·5	37–43
Iso-maltose (%DS)	0·5–2·0	—	—	—	
Maltotriose (%DS)	—	—	15·3	32·7	
Total fermentable sugars	—	—	75·4	101·2	

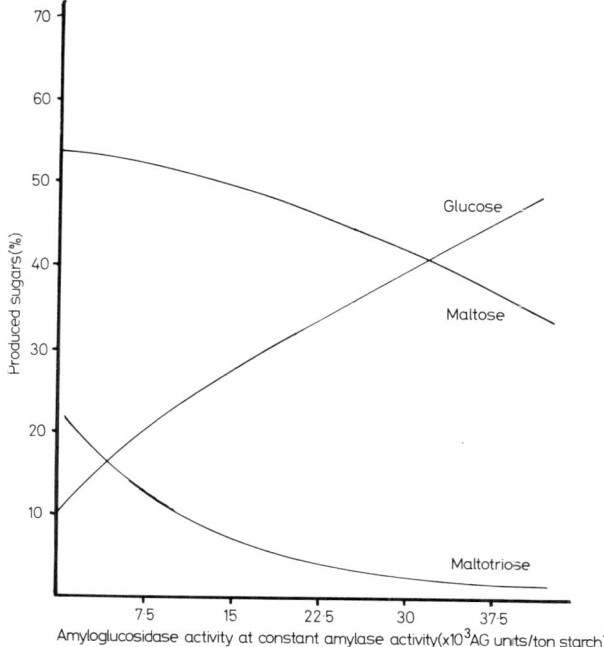

FIG. 8. Production of high conversion syrups from 20 DE acid-converted starch. Data from Reference 60.

commercial scale, using thermostable endo-amylase for liquefaction of starch and glucogenic exo-amylase for saccharification, to give syrups containing > 90% glucose. This development has induced starch processors, enzyme manufacturers and syrup users to enquire into the possibility of producing syrups containing > 60% maltose. However, efficient depolymerisation of starch to its penultimate hydrolysis product, maltose, is more complex than complete hydrolysis to glucose.

The problem

Consider first the hydrolysis of amylose or an α-1, 4- portion of a linear oligosaccharide containing an even number (n) of glucose units. In the presence of a suitable enzyme, $n/2$ mol of maltose would be formed. If the oligosaccharide contained an odd number of glucose units (m), either $(m-1)/2$ mol of maltose + 1 mol of glucose

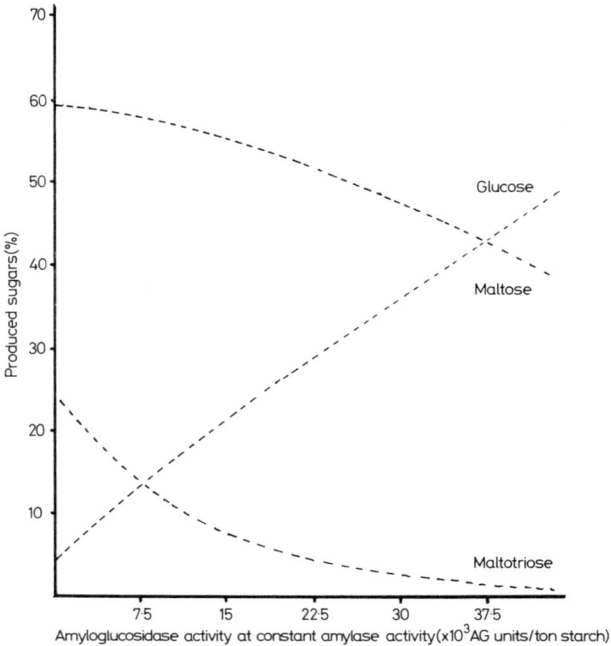

FIG. 9. Production of high conversion syrups from 20 DE enzyme-converted starch. Data from Reference 60.

would be formed, (if a maltogenic exo-amylase, e.g. cereal β-amylase, had been used) or $(m - 3)/2$ mol of maltose + 1 mol of maltotriose (if a thermolabile endo-amylase, e.g. fungal α-amylase, had been used). So this part of the process is straightforward.

The problem becomes apparent when considering hydrolysis of the predominant amylopectin portion of the starch,[61] which contains α-1,6- branch points. This is because maltogenic exo-amylases have higher specificities than their glucogenic counterparts (Table 1) and are unable to by-pass or hydrolyse the α-1,6- links in the same way as amyloglucosidase.[62] Thus cereal β-amylase would begin to hydrolyse off maltose units from the non-reducing ends of the amylopectin-derived branched oligosaccharides stopping within 2–3 glucose units of the α-1,6- branch point, to give β-limit dextrins.[63] Fungal α-amylase, being an endo-amylase, can attack the molecule at either side of the branch point but, being specific for α-1,4- linkages, is unable to hydrolyse the α-1,6- units, and produces smaller α-limit dextrins.[64] Which of the two enzymes is likely to give the highest yield of maltose, will depend upon the position of the α-1,6- branch point (relative to the reducing end of the molecule) and the closeness of approach allowed by the relative specificities of the two enzymes.

Consider the general case of a starch containing a proportion of amylose (a), whose fractional hydrolysis potential to maltose by an enzyme is (x). If the fractional hydrolysis potential to maltose of the amylopectin portion is (y), then the amount of maltose produced by the action of this single enzyme is $ax+(1-a)y$. It has been reported that release of maltose from amylopectin by β-amylase is 50–60% of potential, so that, if it is assumed that the value of (y) is roughly proportional to the average amylopectin chain length, the maximum amount of maltose produced from starch of a given origin can be estimated (Table 17). The maximum amount of maltose which could be envisaged as resulting from the action of either β-amylase and/or fungal amylase must therefore be when

TABLE 17
ESTIMATION OF MALTOSE PRODUCTION USING SINGLE ENZYME PROCESS

Proportion of amylose in native starch = a
Fractional hydrolysis potential to maltose = x
Fractional hydrolysis potential of amylopectin portion = y
Therefore amount of maltose produced by the action of a single enzyme = $ax+(1-a)y$

Starch	Amylose (a)	Amylopectin chain length	Estimated value of y	Maltose (%)
Barley	0·22	23–26	0·56	64·5
Maize	0·24	20–27	0·55	64·6
Wheat	0·25	20–23	0·52	62·8
Potato	0·18	22–24	0·54	61·4
Rice	0·18	22	0·53	60·6
Oats	0·27	20	0·50	62·2

Analytical data from various sources including Reference 66.

$x=1$ and $y \cong 0.6$, i.e. 70% of potential. Thus in order to produce a higher level of maltose, it will be necessary to hydrolyse the α-1,6- linkages, effectively releasing α-1,4- polymers which will now be substrates for the saccharifying enzymes. The effect of such an enzyme on maltose yield can be predicted as follows.

If the number of chains in an amylopectin molecule is (n) and the average number of glucose units per chain is (g), the approximate molecular weight of the amylopectin is given by the expression

$$\text{MW} = ng\,180 + n(g-1)18 + n18$$
$$= 18n(11g)$$

For maize, $n = 215$, $g = 23 \cdot 5$ and MW is estimated as 10^6. Assuming the closest orientation a maltose-producing enzyme can approach an α-1,6- branch point is 2 glucose units, then the proportion of the amylopectin unavailable for maltose production is $6/g \simeq 25\%$. Thus if the α-1,6- linkage was broken, it would be possible to produce a syrup containing $70 + \sim 25 = 90 \rightarrow 95\%$ maltose.

The solution

Such enzymes—termed debranching enzymes—have also been developed during the 1970s and are described in the literature.[61,65] The enzyme which has found commercial application has an original specificity towards the linear α-glucan polymer, pullulan, whose structure is essentially an α-1,6- polymer of maltotriose. This enzyme, termed pullulanase, hydrolyses α-1,6- links at random in pullulan producing hexa and nonaoligosaccharides initially, and maltotriose finally. The specificity is broad enough to accommodate all α-1,6- links in amylopectin and limit dextrins, provided there are at least two α-1,4- glucosidic links on either side of the branch point. Pullulanase from *Klebsiella aerogenes* has reaction characteristics comparable with previously described maltose-producing enzymes and an example of its use is given in Table 16.

More recent examples of the use of pullulanase in producing syrups containing high concentrations of maltose have been reported, one of them at last year's symposium in this series.[55]

The Future

The developments over the past 15 years in the production and application of enzymes have enabled the whole range of syrups and extracts outlined above to be applied on an industrial scale. This branch of biotechnology—design enzyme engineering—has been successful, due not only to the development of specific enzyme systems but equally to the evolution of superb analytical techniques, and realistic process technology (particularly in the areas of reactor design and downstream separation and refining techniques). This has been catalysed by speedy commercialisation and effective marketing by the industries involved, helped by open communication at the earliest stages of development. A stage has now been reached when an enzyme system can be designed to produce a syrup of virtually any specification—high-maltose/low-glucose syrups being a relatively recent example. Future work will involve improving the efficiencies of these processes. More significant developments will be industrial syrup production from carbohydrate sources other than starch—cellulose or lactose, for example—and the production of substances other than sugars—for example, sugar alcohols and sugar-derived organics—on a large scale. What is certain, is that these developments will be equally challenging and that applied enzymology will play a central role.

ACKNOWLEDGEMENT

The author wishes to thank Lidi Villadsen and Laila Jensen for skilled technical assistance, and other former colleagues at Novo Industri A/S, Denmark and Novo Enzyme Products Ltd, England, for assistance and helpful discussion.

REFERENCES

1. ALLEN, W. G. and SPRADLIN, J. E. (1974). *Brewers Digest*, **65**, 48–56.
2. NORMAN, B. E. (1978). *Soc. Gen. Microbial Colloquium*, September, Aberdeen.
3. UNDERKOFLER, L. A. *et al.* (1965). *Die Stärke*, **17**, 179–84.
4. HIGASHIHARA, M. and OKADA, S. (1974). *Agric. Biol. Chem.*, **38**, 1023–29.
5. TAKASAKI, Y. (1976). *Agric. Biol. Chem.*, **40**, 1515–22 and 1523–30.
6. ROBYT, J. and FRENCH, D. (1964). *Arch. Biochem., Biophys.*, **104**, 338–45.
7. NAPIER, E. J. (1977). Brit. P. 4 011 136.
8. KAINUMA, K. *et al.* (1975). *Biochim. Biophys. Acta*, **410**, 333–46.
9. OHBA, R. and UEDA, S. (1975). *Agric. Biol. Chem.*, **39**, 967–72.
10. YOKOBAYASHI, K. *et al.* (1969). *Agric. Biol. Chem.*, **33**, 625–7.
11. STENTEBJERG-OLESEN, B. (1979). *Enzyme Symposium*, October, Shanghai.
12. REESE, E. T. and MANDELS, M. (1966). In: *Methods in Enzymology,*

Vol. 8, Eds Neufeld, E. F. and Ginsburg, V. Academic Press, New York, 104.
13. OUTTRUP, H. (1976). US P. 3 980 521.
14. AUNSTRUP, K. (1974). *Industrial Aspects of Biochemistry*, Ed. B. Spencer. Fed. Europ. Biochem. Socs.
15. SØRENSEN, S. Å. and FULLBROOK, P. (1973). *MBAA Tech. Quarterly*, **9**(4) 166–72.
16. FULLBROOK, P. (1976). *JFST Proceedings*, **9**(3) 105.
17. HARRIS, G. et al. (1955). *Proc. Eur. Brew. Conv.*, Baden-Baden, 34.
18. *Novo Enzyme Information*. (1971). EB 9–15 August, Novo Industri A/S 2880 Denmark, 25.
19. DE DEKEN, R. H. (1966). *J. gen. Microbiol.*, **44**, 149.
20. HARRIS, G. (1962). *Barley and Malt*, Ed. Cook, A. H. Academic Press, London, 583.
21. HOPKINS, R. H. and KRAUSE, B. (1947). *Biochemistry Applied to Malting and Brewing*, George Allen and Unwin, London.
22. NORRIS, F. W. and DARBRE, A. (1956). *Analyst*, **81**, 394.
23. HOPKINS, R. H. and WEINER, S. (1944). *J. Inst. Brew.*, **50**, 124.
24. TULLO, J. W. and STRINGER, W. J. (1945). *J. Inst. Brew.*, **51**, 86.
25. HOPKINS, R. H. and WEINER, S. (1945). *J. Inst. Brew.*, **51**, 34.
26. BOLLINDER, A. et al. (1956). *J. Inst. Brew.*, **62**, 497.
27. NORRIS, F. W. (1945). *J. Inst. Brew.*, **51**, 177.
28. HOPKINS, R. H. et al. (1948). *J. Inst. Brew.*, **54**, 264.
29. HOPKINS, R. H. and PENNINGTON, R. J. (1947). *J. Inst. Brew.*, **53**, 251.
30. LYNES, K. J. and NORRIS, F. W. (1948). *J. Inst. Brew.*, **54**, 150, 207.
31. JONES, M. and PIERCE, J. S. (1963). *Proc. Eur. Brew Conv.*, Brussels, 101.
32. JACKSON, S. W. and HUDSON, J. R. (1978). *J. Inst. Brew.*, **84**, 34.
33. BARNES, H. M. and KAUFMAN, C. W. (1947). *Ind. Eng. Chem.*, **39**(9) 167–70.
34. HARTLEY, D. (1955). *Food in England*, MacDonald and Jane's, London, 540.
35. PREECE, I. A. and MACDOUGALL, M. (1958). *J. Inst. Brew.*, **64**, 489.
36. HOUGH, J. S. et al. (1971). *Malting and Brewing Science*, Chapman & Hall, London, 153.
37. SØRENSEN, S. Å. (1970). *Process Biochem.*, **5**, 60.
38. SØRENSEN, S. Å. (1971). *Affinidat*, XXVIII, 651.
39. SØRENSEN, S. Å. (1973). *Der Brauer und Mälzer*, August 27.
40. STENTEBJERG-OLESEN, B. (1971). *Process Biochem*, **6**, 29.
41. WIEG, A. J. et al. (1969). *Process Biochem.*, **4**, 33.
42. WIEG, A. J. (1970). *Process Biochem.*, **5**, 46.
43. HARRIS, J. O. (1968). *Brew. Guard.*, May, 81.
44. KLOPPER, W. J. (1968). *Int. Tijdschrift voor Brouwenjen Moutenj*, **5**, 143.
45. LATIMER, R. A. et al. (1966). *Inst. Brew. (Aust.) Proc. Conf.*, **9**, 111.
46. MACEY, A. et al. (1967). *Proc. Eur. Brew. Conv.*, Madrid, 283.
47. HANSEN, M. (1972). *Brygmesteren*, **3**, 65.
48. NIELSEN, E. B. (1971). *Proc. Eur. Brew. Conv.*, Estoril, 149.

49. ENEVOLOSEN, B. S. (1970). *J. Inst. Brew.*, **76**, 546.
50. ENEVOLOSEN, B. A. (1971). *Brygmesteren*, **2**, 41.
51. BLEY, W. (1970). *Intn. Brewers J.*, (Jan), 328.
52. SØRENSEN, S. Å. (1973). *Der Brauer und Mälzer*, (Aug).
53. SCHWIMMER, S. (1947). *Cereal Chem.*, **24**, 167.
54. ERLICH, V. L. and BURHERT, G. M. (1949). *Cereal Chem.*, **26**, 326.
55. TAKASAKI, Y. and YAMANOBE, T. (1981). In: *Enzymes and Food Processing*, Eds Birch, G. G., Blackebrough, N. and Parker, K. J. Applied Science Publishers, London, 73.
56. BARFOED, H. C. (1976). *Cereal Foods World*, **21**, 588.
57. FULLBROOK, P. D., VABØ, B. and ØSTEGAARD, J. (1977). *Ind. Chem. Eng. Symposium No 51*, London.
58. LÜTZEN, N. W. (1981). *Novo Industri A/S House Journal*, NP80.
59. ALLEN, W. G. and SPRADLIN, J. E. (1974). *Brewers Digest* (July) **48**, 65.
60. MADSEN, G. B. and NORMAN, B. E. (1973). In: *Molecular Structure and Function of Food Carbohydrates*, Eds Birch, G. G. and Green, L. F. Applied Science Publishers, London, 50.
61. BANKS, W. and GREENWOOD, C. T. (1975). *Starch and its Compounds*, Edinburgh Univ Press.
62. PAZUR, J. H. and ANDO, T. (1960). *J. Biol. Chem.*, **235**, 297.
63. ROBYT, J. F. and WHELAN, W. J. (1968). In: *Starch and its Derivatives*, 4th edn, Ed. Radley, J. A. Chapman & Hall, London, 477.
64. ROBYT, J. F. and WHELAN, W. J. (1968). In: *Starch and its Derivatives*, 4th edn, Ed. Radley, J. A. Chapman & Hall, London, 430.
65. ALLEN, W. G. and DAWSON, H. G. (1975). *Food Technol.*, (May), 70.
66. WILLIAMS, J. M. (1968). In: *Starch and its Derivatives*, 4th edn, Ed. Radley, J. A. Chapman & Hall, London.

5

Nutritive Sweeteners from Starch

T. J. PALMER
*Tunnel Refineries Ltd,
Greenwich, London, UK*

ABSTRACT

In modern industrial food and drink production, the technical characteristics of glucose syrups/dextrose make them an essential ingredient in recipes enabling modern plants and processes to work efficiently and consumers' requirements to be met. The physico-chemical characteristics of glucose syrups/dextrose are distinct and valuable. Glucose syrups can be manufactured and blended for incorporation in a food recipe such that any of the following technical characteristics are utilised:

1. *range of moisture retention*
2. *range of viscosity*
3. *range of crystallisation inhibition (particularly preventing sucrose crystallisation in a wide range of products)*
4. *range of freezing point control*
5. *range of sweetness control (allowing improved flavour transfer)*
6. *improvement in appearance (gloss, lustre, clarity)*
7. *greater preserving action by lower molecule weight*
8. *foam stabilisation and binding action*
9. *range of fermentation control*

Glucose is by far the most widely distributed sugar in nature. It is the building block for all starch and cellulose. Humans and other mammals were metabolising starch long before sucrose was discovered. The processes used in the glucose syrup industry have been modelled from the natural conversion of starch, through *intermediate dextrins and sugars, to dextrose (glucose or blood sugar). It is natural that the most commonly occurring sugar unit, glucose (dextrose), is essential for sustaining human life. Glucose is our most basic and valuable energy source.*

Nutritive sweeteners from starch have been used for well over a century. This paper will cover the developments in manufacture, properties and applications of traditional glucose syrups, dextrose and high fructose glucose syrups.

INTRODUCTION

Sources of starch and industrial extraction have been summarised by Knight.[1] Starch is a constantly replenished carbohydrate, being the natural energy store of a large variety of plants. Plants are able to convert water, carbon dioxide and energy from sunlight into simple water soluble sugars (e.g. glucose). Subsequently these sugars are built into ever more complex polymeric, insoluble structures. The result is an aggregate of semi-ordered, polysaccharide molecules called a starch granule, built entirely of glucose-type units. The bio-synthesis and physico-chemical structure of starch have been exhaustively summarised by Banks and Greenwood.[2] Industrial conversion of starch (by acid or enzyme hydrolysis) is aimed at unravelling this complex polymer.[3] The potential range of glucose syrup products is very wide and may be tailor-made for specific applications.

A historical summary of advances in the production of sweeteners from starch shows the following:

1850s — acid converted glucose syrup

1920s — crystalline dextrose

Late 1930s — acid–enzyme converted glucose syrups

Late 1960s — multiple enzyme dextrose and high fructose conversion

Thus starch is the very flexible commercial source of a 'total range of sweeteners' (exceptions being sucrose and lactose). This is indicated in Table 1.

TABLE 1
CARBOHYDRATE COMPOSITION OF SYRUPS FROM STARCH

	Glucose	High maltose	Sweetose®	Dexyme®	Isosweet®
Dextrose Equivalent	42	42	63	98	98
Dextrose (% DS)	19	3	39	94	52
Maltose (% DS)	14	50	33	3	3
Maltotriose (% DS)	11	10	9	1	1
Higher sugars (% DS)	56	37	19	2	2
Fructose (% DS)	—	—	—	—	42

® Registered Trade Marks of Tunnel Refineries Ltd., London.

GLUCOSE SYRUPS

Definition

Glucose syrup is a food and may be further defined as the purified concentrated aqueous solution of nutritive saccharides obtained from starch, having a DE of 20 or more. It is also sold in a spray-dried form.

Syrups are identified by reference to the DE value. The abbreviation DE stands for 'dextrose equivalent', which refers to the measurement of the total reducing sugars in the syrup, calculated as dextrose (D-glucose) and expressed as a percentage of the total dry substance of the solution. Thus starch has a DE of 0 and dextrose has a DE of 100.

Starch is a giant molecule composed of D-glucose units. Total hydrolysis of the starch molecule would thus yield only D-glucose (dextrose). Traditional glucose syrups are sweeteners made by less than total hydrolysis of starch with controlled DE between 20 and 98. This hydrolysis is brought about by acid or enzymes or both.

Manufacture

The manufacture of glucose syrup is a multi-step process. Starch is converted to a low solids syrup or liquor which is then refined and concentrated to commercial solids concentration. Three common procedures are used. They are: (a) the acid conversion process, (b) the acid-enzyme conversion process, and (c) the multiple enzyme process. In the acid conversion process, a starch slurry is acidified to about pH 2·0 and pumped into a continuous reactor which operates at elevated temperature and pressure. After the proper time interval, the liquor is returned to atmospheric conditions and neutralised. The liquor is clarified, decolourised with activated carbon and concentrated by evaporation to finished solids.[3] The acid process is generally used for syrups in the range 30–55 DE.

The acid–enzyme process is similar. The acid conversion is completed and then one or more saccharifying enzymes are added to produce higher DE syrups.[4] For example, in the production of 42 DE high maltose syrup, the acid conversion is halted at the completion of starch solubilisation. A maltose-producing saccharifying enzyme, e.g. malt extract, is added and the conversion continued until the desired level of maltose is produced (in the range 35–55%). In the case of 63 DE glucose syrups, both maltose-producing and dextrose-producing enzymes are added and the reaction stopped when both saccharides are in the 33–40% range.[3,4] The enzymes are then deactivated and the liquor is clarified, decolourised and evaporated.

Multiple enzyme processes are used for those syrups with the highest level of maltose and glucose content, e.g. 95%+ dextrose. Dried solids are usually made from the lower DE syrups by spray or drum drying. Because of hygroscopicity, they must be packed in moisture-barrier bags.

General Characteristics and Properties

It should be noted that acid conversion always yields a product of specific composition because of the random hydrolysis of the starch. In contrast the combination of acid and enzyme processing results in a syrup whose composition depends upon the acid conversion step as well as upon the specific enzyme used.[5]

In addition to describing the syrup by DE alone, it is frequently useful to state the composition in terms of DE and one or more of the saccharide fractions. A good example is the 42 DE high maltose syrup made by an acid–enzyme process. In this case, a partially acid converted glucose syrup is treated with enzyme to produce a syrup with about 45% maltose and 6% dextrose.[3]

Solids content

Glucose syrups are also characterised according to solids content. Most commercial syrups are sold on a Baumé (Bé) basis, which is a measure of the dry substance content by specific gravity and represents about one half of the solids content. However, in production, Baumé determination has been largely superseded by the direct determination of solids by refractive index.[5] Most glucose syrups are available in the range 39 to 45 Baumé, which corresponds to a dry substance content of about 71 to 85%.

pH values

Glucose syrups are finished and stored slightly on the acid side. They are available in a pH range of about 4·5 to 5·5.

Sweetness

Glucose syrups have a clean, pleasant, sweet taste and the degree of sweetness is dependent upon the concentration and the type of saccharides.[6] On a relative scale of sweetness, fructose is sweeter than dextrose which in turn is sweeter than maltose, and maltose is sweeter than the higher saccharides. Since the mono- and disaccharides also have a higher reducing power (DE), the DE value of glucose syrups provides some indication of relative sweetness. In general, the higher the DE, the sweeter the syrup.[4, 7]

When glucose syrup is used in combination with sucrose, the resulting sweetness is usually greater than expected. For example, when tested at 45% solids, a mixture of 25% 42 DE glucose syrup and 75% sucrose is considered to be as sweet as a sucrose solution of 45% solids. Hence, when glucose syrups are used with sucrose in products having a relatively high total sweetener concentration, no apparent loss of sweetness results, e.g. in toffees or high boiled sweets and jams.

Sweetness is also influenced by temperature, acids, salts, flavouring materials, etc. It is difficult, therefore, to state quantitatively the relative sweetness of syrups and sugars for use in food formulations.[7] Each formulated product needs to be considered individually.

Hygroscopicity

Glucose syrups are hygroscopic to varying degrees. The extent of hygroscopicity increases as the DE increases since hydration is related chiefly to the monosaccharide content. Glucose syrups and dextrose are employed as moisture retention aids, crystallisation inhibitors, stabilisers, etc.

Textural characteristics

The higher saccharides give glucose syrup its cohesive and adhesive properties. They also contribute a chewy texture character to certain types of confections, e.g. toffees and chewing gum.

Blends of nutritive sweeteners offer flexibility to formulators in controlling crystallisation of certain sugars and improving shelf-life characteristics. This property is advantageous in many types of sweet products such as ice cream, jams, jellies, preserves and confections.

Molecular properties

Dextrose and fructose are monosaccharides with a molecular weight of 180 (compared with 342 for sucrose) and have a relatively high osmotic pressure which increases their effectiveness in inhibiting microbial spoilage. They can be used to reduce water activity without contributing excess sweetness. Glucose syrup of 55 DE has about the same average molecular weight as sucrose or lactose and hence, has about the same osmotic properties as these sugars. Glucose syrups of lower DE have higher molecular weights with correspondingly lower osmotic pressures.

The effect of glucose syrups and sugars on the freezing points of solutions is of practical significance in the manufacture of ice cream and frozen desserts. Control of the freezing point is obtained by blending sucrose, high fructose glucose syrups and glucose syrup of the appropriate DE.[5,7]

Viscosity

Viscosity, one of the most important physical properties of glucose syrups, is dependent on density, DE and temperature. Viscosity decreases as DE and temperature are raised but increases with higher density.

Fermentability

Fermentability is an important property of glucose syrups particularly in the baking and brewing industries. The lower molecular weight sugars, mainly the mono- and disaccharides, glucose, fructose and maltose, are readily fermentable by yeast. The total fermentability of glucose syrups is roughly proportional to their content of mono-, di- and trisaccharides and the higher the DE the higher the fermentability.[8]

Reducing characteristics

Dextrose and fructose produce browning which can be desirable in baked and other goods.

The individual component sugars of corn sweeteners are technically termed 'reducing sugars'. Their 'reducing' property is useful in maintaining the bright red colour of tomato ketchup and strawberry preserves. This reducing action of syrups and sugars is also measured and expressed by DE.[5]

APPLICATIONS OF GLUCOSE SYRUPS

Confectionery

It is of paramount importance in any sugar confection to control the tendency of sucrose to crystallise. The simplest and most efficient way of doing this is by incorporating some type of glucose syrup in the recipe. Boiled sweets are unlikely to have more than 2% moisture in them, as their physical nature is that of a supercooled solution (i.e. similar to glass). Moisture adversely affects any crystal clear boiling. A minute layer of water on the surface of the sweet will lead to that progressive deterioration in quality called 'shell-graining'. This is caused by local surface dilution of the sweet causing the sucrose to crystallise out . The dilution layer moves towards the centre of the sweet leaving concentric layers of crystallised sucrose.

When sucrose is boiled it breaks down to its constituents—fructose and dextrose, both reducing sugars. This phenomenon (process inversion) produces increasingly more reducing sugars with time. Sugar boiling in open atmospheric pans gives rise to far more reducing sugars than the modern vacuum pan or microfilm continuous cooker operations, where process inversion is negligible. A certain content of reducing sugars is needed to control shell-graining; too much process inversion makes the sweet very hygroscopic, i.e. the sweet 'runs to syrup' on exposure. The incorporation of glucose syrup (42 DE acid converted or high maltose) in a recipe makes the task of balancing the final product composition easy, especially in modern minimum inversion systems. The amount of reducing sugars in a glucose syrup does not increase during processing. Therefore the reducing sugars in the boiling can be controlled within close limits by measuring the required quantities of glucose syrup (known reducing sugar content) and sucrose into the premix machine.[7]

Glucose syrup reduces the effective concentration of sucrose in a high boiling, hence reducing the tendency of sucrose to crystallise. Also, on exposure to moisture and dilution of the sweet surface, the glucose syrup formed is very viscous and reduces the rate of migration of the dilution layer. Therefore glucose syrup in the recipe reduces the tendency for crystallisation, and also slows down the rate of shell-graining should moisture affect the boiling. The syrup also contributes to the body and 'mouth-feel' of the confection on consumption. These basic facts also hold good for the incorporation of glucose syrup in toffees, i.e. glucose syrup prevents sucrose granulation and gives the toffee a smooth chewy mouth-feel. Glucose syrup in fudge or fondant controls granulation and produces very fine crystals giving a smooth, short texture without any tendency to be 'gritty'.

Starch jellies and gum drops are cooked mixtures of sucrose and glucose syrup to which starch has been added. Again the correct combination of the three products is essential for proper body, plasticity and mouth-feel. The 63 DE glucose syrups are particularly useful in products with a tendency to dry out, e.g. marshmallows and soft centres.

Jams and Jellies

The prime function of sugars in jam is to preserve the fruit for later consumption. The concentration of sugars in the final jam produces a high osmotic pressure. Any yeast (or other organism) contaminating the jam is dehydrated and rendered unable to metabolise. The jam therefore remains non-degraded and edible. The choice of sugar to produce this high osmotic pressure is not too critical (though dextrose and fructose have a greater effect than sucrose, maltose and higher molecular weight sugars). A high final solids concentration is however most important in this function.

On storage, if the process inversion is low, jam may deposit sucrose crystals. By incorporating glucose syrup (e.g. 63 DE) in the recipe, sucrose crystallisation will not occur. Modern vacuum evaporator systems for jam manufacture need glucose syrup of this type in the formula, to compensate for the inherent low process inversion. The slightly reduced sweetness of jam containing glucose syrup lets the flavour of the fruit develop more fully as the palate is not saturated by sweetness alone. This enhances the satisfaction of the consumer.[7]

Glucose syrups are also used in tablet jellies to prevent sucrose crystallisation and thus preserve the clarity and appearance of tablet packs during transport and storage. A tablet showing evidence of sucrose crystallisation on its surface is unsaleable.

Soft Drinks

Glucose syrups used for soft drink formulations range from the relatively simple acid–enzyme converted types to the multiple enzyme converted, deionised types. These syrups are supplied to specifications set in the range 71 to 80% solids which contributes to their excellent microbiological stability.

The 63 and 98 DE types when tasted by themselves are less sweet than sucrose. But, depending upon the final products and particular recipe, these glucose syrups may form up to 30% by weight of the carbohydrate sweetener in a soft drink formulation. Mixtures of glucose syrups and sucrose appear to be sweeter than the individual components tasted separately. Also certain glucose syrups appear to have a beneficial effect in formulations containing saccharin, especially in citrus drinks for dilution before consumption.[9] This effect is two-fold, the glucose appears to gain sweetness from the saccharin, while the bitter 'after-taste' of saccharin appears to be masked by the presence of the glucose syrup.

Canning

A canned food should closely resemble the original fresh produce. Glucose syrups are used to: (a) produce a bodying effect in the cover syrup medium; (b) balance the sweetness of the produce (especially useful where the natural sweetness increases as the season advances and the produce ripens[5]); (c) provide sufficient osmotic pressure to preserve the produce without cloying sweetness (a 34 DE low sweetness, glucose syrup is especially useful in this respect for canned vegetables); (d) give a quicker and better penetration of the fruit (by the incorporation of 63 DE glucose syrup) making the fruit more succulent.[7]

Baking

The use of 63 DE glucose syrups or dextrose monohydrate can have a beneficial effect in the manufacture of bread, buns, and cakes. The humectant properties can help to retain moisture in products under widely varying conditions thus preventing staleness or drying out. Protection against moisture loss is particularly important to the keeping qualities of sponge-type cakes. The dextrose and maltose contained in these products is directly fermentable by yeast and therefore dough fermentation can be swiftly and efficiently accomplished.[7] On baking, the glucose syrup can caramelise readily and produce an attractive crust colour for bread: easy caramelisation is also important in dark fruit-cake. Glucose syrups can also contribute to the body and gloss of fruit pie fillings.[5] Fondants and butter creams for inclusion in bakery products are improved considerably when the total formulation contains 63 DE glucose syrup or dextrose monohydrate.

Fermentation

The fermentation industries, i.e. brewing, cider manufacture and winemaking, utilise the readily fermentable sugars in glucose syrups for production of alcohol.[8,10] For high fermentability, i.e. maximum conversion of carbohydrate to alcohol, the 98 DE (high dextrose deionised) syrups are especially valuable for cider and wine fermentation. The 63 DE (dextrose/maltose) syrups are a closer match to the sugars and dextrins produced from brewers malt during the mashing process. The complete carbohydrate spectrum of these syrups contributes alcohol, potential flavour components, and control of other desirable properties of the finished beer. In particular, high gravity fermentations benefit from these cereal syrups.[10]

Dietetic and Infant Foods

Glucose syrup is added to health drinks to help convalescence. The syrup provides a convenient and instant source of energy requiring little digestion. Certain very high calorie formulae based on glucose syrups are being successfully used in the treatment of some acute liver complaints.[11] Glucose syrups are also widely used in baby syrups and foods to provide a balanced carbohydrate source, i.e. a mixture of instantly available sugars plus higher molecular weight carbohydrates for later conversion in the body.

Ice Creams

Ice cream formulations contain glucose syrups ranging from the sweetest to the very low (20 DE) malto-dextrin types. Glucose syrups in ice cream function in a variety of ways: they provide better melt-down characteristics, serve as crystallisation inhibitors, give better freezing-point control, contribute to the body and mouth-feel of the product and provide a balanced sweetness.[5]

DEXTROSE

Introduction

The monosaccharides and amino acids are the most common organic chemical compounds found in nature. The monosaccharide, glucose, is by far the most abundant and without it, most of the earth's plants and animals could no longer exist. Cellulose, starch and glycogen are composed almost entirely of anhydroglucose molecules linked together into long chain polymeric forms. Upon hydrolysis, the naturally occurring disaccharides, e.g. lactose and sucrose, yield one glucose molecule and one molecule of an isomeric monosaccharide, i.e. galactose and fructose respectively. Consequently, it follows that man's diet includes glucose in rather substantial quantities. In both combined and free states it comprises the major part of all carbohydrate in the diet.[5, 12]

It is useful here to define the two terms 'glucose' and 'dextrose'. Glucose is the correct chemical term used to identify a specific aldohexose with the formula $C_6H_{12}O_6$. Dextrose is the commercial designation for α-D-glucose generally as the monohydrate.

Dextrose monohydrate is purified and crystallised D-glucose containing one molecule of water of crystallisation with each molecule of D-glucose. The moisture content is not more than 10% (w/w basis) and the reducing sugar content (dextrose equivalent) not less than 99·5% (w/w dry basis). The sulphated ash content is not more than 0·10% (w/w dry basis) and sulphur dioxide content not more than 15 ppm (w/w dry basis).

Physical and Chemical Properties of Dextrose

Crystalline dextrose is a white solid which exists in three isomeric forms, anhydrous α-D-glucose, α-D-glucose monohydrate and anhydrous β-D-glucose. Physical properties of these isomers are summarised in Table 2.[18]

TABLE 2
PHYSICAL PROPERTIES OF D-GLUCOSE

	α-D-Glucose	α-D-Glucose hydrate	β-D-Glucose
Formula	$C_6H_{12}O_6$	$C_6H_{12}O_6 \cdot H_2O$	$C_6H_{12}O_6$
Melting point (°C)	146	83	150
Solubility at 25°C (% by wt)	62	30.2	72
Optical rotation $[\alpha]_D^{20}$	112.2–52.7[a]	112.1–52.7[a]	18.7–51.7[a]
Heat of solution at 25°C (cal/g)	−14.2	−25.2	−6.2

[a] Rotation of alpha/beta equilibrium mixture.

The crystalline forms of dextrose are either α-D-glucose monohydrate or anhydrous α-D-glucose. The β-form is unstable and exists only in solution unless special effort is made to isolate it, e.g. by precipitation from a pyridine solution. If pure β-D-glucose is dissolved in water it will undergo mutarotation to an equilibrium mixture of about 62% of the β-form and 38% of the α-form at 25°C and 30.2% solids content. If that solution is concentrated, the less soluble form crystallises out as α-D-glucose monohydrate until the solids content is again lowered to 30.2%. The equilibrium, i.e. 62% β-, 38% α- will be maintained by continuing mutarotation.[5]

Chemical Reactivity

Dextrose is an aldehyde and as such exhibits many, but not all, the typical aldehydic reactions. It is also a primary alcohol, a secondary alcohol and a polyhydric alcohol and, under the proper conditions, will react as a member of these classes of compounds. All these properties may be used to advantage in the commercial applications for dextrose in which it is a chemical raw material or intermediate rather than a food ingredient.

Dextrose is a reducing sugar and will reduce copper in Fehling's solution. This reducing property makes possible the classical detection and assay procedures used for dextrose and dextrose-containing materials. It also makes dextrose useful as a reagent, e.g. in leather tanning.[5]

When dextrose is dissolved in water it absorbs heat. This property too, is useful in certain specialised applications, e.g. in confections in which a 'cooling' effect is desirable, or in baking where moderating dough temperature is an important consideration in yeast-raised products.

Dextrose Digestion and Utilisation

Dextrose is a pure, highly nutritious food and is useful and advantageous to the food industry, being a very economical source of pure carbohydrate.

Dextrose is the common intermediary metabolite in carbohydrate metabolism, since other utilisable monosaccharides are converted to dextrose before they are further metabolised. Starch, glycogen, and the common disaccharides are hydrolysed enzymatically in the alimentary canal. Dextrose is normally absorbed into the portal vein blood by which it is transported first to the liver and from this organ through the general circulation to all parts of the body. Before glucose or any other monosaccharide can be utilised metabolically, they must be phosphorylated and enter the glycolytic cycle. Other monosaccharides, e.g. fructose, sooner or later are converted to glucose-6-phosphate, in which form they lose specific identity and beyond which they are metabolised like dextrose.[5] In the glycolytic cycle ingested dextrose (or glucose) is temporarily converted to the polysaccharide glycogen or 'animal starch' for storage, for example, in the muscles and liver.

Upon demand by the body tissues, because of physical or mental exertion, the glycolytic cycle reverses and stored glycogen is hydrolysed back to glucose-1-phosphate and from that to the utilisation reaction chain. This is a quite complex, multi-step process which results in the release of life-sustaining energy, the return to the blood stream of oxidation end products (carbon dioxide and water) and their eventual expulsion.[13]

The caloric value of dextrose is approximately 4 cal/g dry substance. Whether consumed as the monosaccharide itself or as a component of fruits, berries, starch, starchy foods, vegetables or glycogen in muscle meats and liver, the compound is quickly and completely absorbed in the mouth, stomach and small intestine. In as little as 30 min after a meal, human blood glucose will be noticeably elevated from its average normal concentration, indicating the rapidity with which the compound is assimilated.

Manufacture

Dextrose was first manufactured successfully on a commercial basis in the USA during the early 1920s by crystallisation from a starch acid hydrolysate solution. The most modern and energy efficient systems of dextrose production involve starch being gelatinised and thinned by pasting and multiple enzyme systems. The enzymes most commonly used for saccharification are glucoamylases of fungal origin that are specific for the production of dextrose.[15,18]

The starch hydrolysate, which takes 48–96 h at about 60°C to produce, is then filtered and centrifuged to remove remaining traces of insoluble lipids, protein and starch. The filtrate is refined with granular or powdered activated carbon plus ion-exchange resins to remove colour and trace impurities, such as soluble inorganic salts and proteinaceous residues. The refined, decolourised liquid is concentrated to about 75% solids content by multi-stage vacuum evaporation and cooled to about 45°C. It may then take one of three routes. It may be sold as a finished syrup containing about 95% dextrose; it may be fed as the starting liquor to an isomerisation plant to be made into high fructose glucose syrup; or it may be fed to a crystallisation plant to be made into dextrose.

In the crystallisation plant the refined liquor is further cooled and then fed to crystallisers, which are large jacketed vessels with slow moving agitators already containing seed crystals. These vessels are kept cool and the refined liquor is gradually lowered in temperature over a period of several days to about 25°C. At that point some 60% of the dextrose originally present has formed crystals and the mixture of liquor and crystals is called 'magma' or 'massecuite'.

The process from that point on is very similar to that of a sucrose refinery. The magma is pumped to centrifuges, the liquid phase (greens) is spun off and the crystals are washed with a spray of steam/water. The damp dextrose cake containing about 14 to 15% moisture may then be treated in one of two ways. It may be dropped to a remelt tank to be made into liquid dextrose (a 71% solids solution) or air-dried to about 8·5% moisture (i.e. 0·6% below the theoretical moisture content for dextrose monohydrate). The whole manufacturing process is summarised in Fig. 1.

The liquor spun off in the centrifuges is still, of course, quite rich in dextrose. This 'first greens' is again concentrated by evaporation

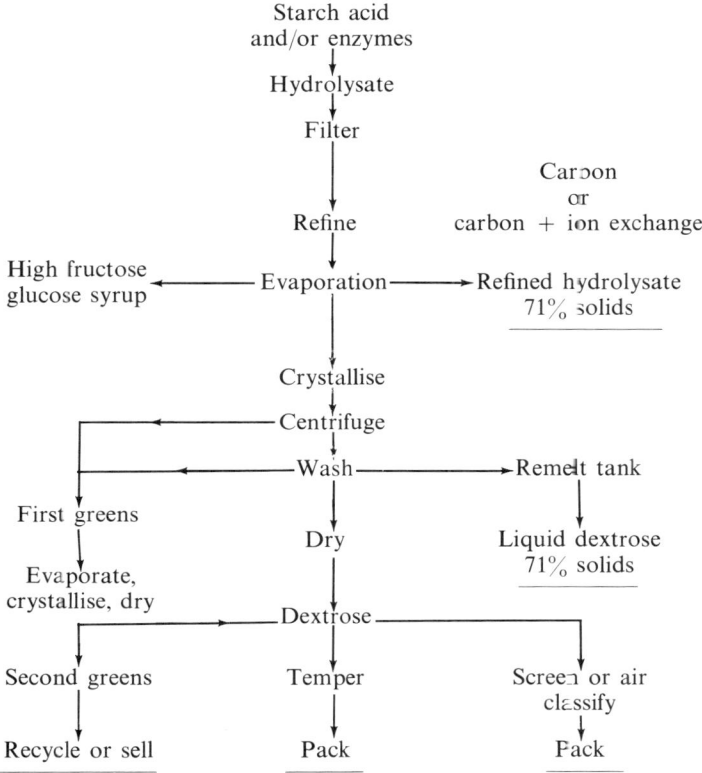

FIG. 1. Manufacture of dextrose.

to 75% solids, cooled, fed back to other crystallisers and the whole process repeated to provide a second 'crop' of crystalline dextrose. In some processes a third such 'crop' is possible using the concentrated 'second greens'. Depending on the process used, 'greens' is eventually sold as a fermentation substrate or as the starting material for caramel colour and cattle lick blocks.

The dried crystalline dextrose is transferred to tempering bins and packed into paper bags or bulk tankers. It may go from the drier to be separated into coarse and fine crystalline forms[15] before final packing.

Anhydrous dextrose is produced by evaporative crystallisation of redissolved dextrose monohydrate in a vacuum pan. The temperature of crystallisation is regulated at 60 to 65°C, the process takes place in 5 to 6 h and yields are about 50%. The spent mother liquor is recycled to the monohydrate crystallisers.

APPLICATIONS OF DEXTROSE

Baking

About one-third of all the dextrose sold in the USA is used in baked goods, most of it in bread, buns and rolls.[5] In yeast-raised products dextrose serves three purposes: it is a highly efficient fermentation substrate for yeast; it is a moderate sweetener; and it enhances crust colour, texture, crumb colour and crumb character.

It is used in doughnuts, cakes and biscuits as a sweetener, to aid in browning and to lengthen shelf-life. It is also used extensively in icings, for dusting purposes, e.g. for doughnuts or swiss rolls, and as an ingredient in creme fillings.

Confectionery

Another principal use for dextrose is in tabletted confections and chewing gum. It is used for its sweet but not overwhelming flavour; it provides a cooling effect because of its negative heat of solution,[15] and it serves as a humectant to extend shelf-life.

Anhydrous dextrose is used in chocolate products. In the USA this use of dextrose is limited in chocolate by a standard of identity to not more than one-third the total amount of the sweetener constituents present. This standard of identity for chocolate is the only existing standard that limits the use of dextrose in the USA.

Canning and Packing

With the advent of high fructose glucose syrup, the use of dextrose in the canning and packing of fruits, vegetables and prepared foods and drinks, has declined sharply, but substantial quantities are still used, particularly in sauces, soups, gravies, citrus fruits and juices.

Prepared Dry Mixes

Dextrose is valuable in prepared mixes, i.e. dough baked goods such as cakes, rolls, biscuits, doughnuts, pancakes and even speciality breads. It is used in these products for the same reasons, described earlier, as it is used in commercially prepared baked goods. It is also an ingredient in icing mixes in which it is used along with glucose syrup solids to temper the intense sweetness of sucrose and for its humectant properties. It is also used as a carrier for instant drinks, flavours, spices, etc.

Dairy Products and Frozen Desserts

Ten years ago large quantities of dextrose were used in ice cream, sherbets and frozen speciality items, e.g. 'popsicles', ice milk and imitation ice cream. Substantial amounts are still used in 'popsicles' and speciality desserts but the major uses, ice cream and sherbets, have been heavily eroded by a switch to conventional glucose syrup and high fructose glucose syrup.

Pickles

Pickled products account for a significant part of the dextrose used by canners and packers. Depending on the kind of pickle, 25 to 35% of the pickling liquor may be made up of dextrose. Because of the very high osmotic pressure of this monosaccharide in solution the pickling liquor penetrates more rapidly causing less shrinkage and quicker size recovery.

Beer and Wine

Steadily growing quantities of dextrose are being used in the USA brewing industry for the production of the so-called 'low calorie' beers and to a lesser extent in other speciality items made by the brewing industry. Much, if not all, this use is via liquid dextrose products which are more economical to handle in large quantities.[7,8] Dextrose is useful in this application because of the ease and rapidity with which it is fermented by yeast.

Where permitted in the manufacture of wine, dextrose is used to adjust the sugar content and acidity of grape juice, a process called 'amelioration'.

Meat Products

Dextrose is used in a wide variety of meat products, such as sausage, meat loaf, luncheon meats, chopped ham, pressed ham and corned beef hash. It is used for its flavour-enhancing properties and as a carrier for spice oleoresins.

Soft Drinks and Thirst Quenchers

Dextrose is frequently used as a sweetener in fruit juices, still drinks and in the class of drinks called 'thirst quenchers', which can also incorporate certain mineral ingredients intended to maintain the mineral balances in the body during periods of heavy exertion and copious perspiration.

Drugs and Chemicals

Dextrose usage in drugs and chemicals is the second largest in volume for the product but in a wide diversity of applications, no single one of which is very large. One of the largest is as a starting material for the manufacture of sorbitol and the related compound, ascorbic acid.

Dextrose is also used: (a) in the manufacture of citric acid and hence the various commercial citrates; (b) in the manufacture of gluconic and glucoheptonic acids which are sequestering agents; and (c) as an aid in the manufacture of various pharmaceutical products which are made by fermentation processes, such as penicillin and other antibiotics. Itaconic acid for the production of man-made fibre is produced by a similar fermentation.

In many of these applications bulk liquid delivery of deionised crystalliser feedstock is the preferred commercial input.[7]

HIGH FRUCTOSE GLUCOSE SYRUPS

Introduction

Glucose syrups of varying sweetness have been established food ingredients for many years; however fructose was not a component of these syrups until the late 1960s. At this time commercial syrups containing 15 to 42% fructose were developed and described as 'high fructose corn syrup' (HFCS) in the USA, as 'high fructose glucose syrup' (HFGS) in Europe and as 'isoglucose' latterly within EEC legislation.

Typical Analysis—42% Fructose: Isosweet®

Solids 71%
Colour (CIRF × 100) 0·4

®Registered Trade Mark of Tunnel Refineries Ltd., London.

Ash (sulphated), DS	0·03%
Fructose, DS	42%
Dextrose, DS	52%
Other saccharides, DS	6%

Glucose Isomerase Enzyme

For many years the wet-milling industry had directed research efforts towards the development of sweeter corn syrups containing fructose. Attempts through alkaline isomerisation resulted in products with colour, flavour and compositional problems. Compare this with the early manufacture of dextrose monohydrate from high DE acid converted feedstocks.[7] An alternative approach of isomerising glucose to fructose using enzymes was investigated in Japan, the USA and Europe.

To develop an economical process, researchers concentrated their efforts on increasing the yields of glucose isomerase. These studies were successful and resulted in the first commercial process that used a soluble enzyme to isomerise a portion of the glucose to fructose.

The technique of immobilising enzymes on whole cells or on other supports, such as cellulose, offered a further opportunity for improving efficiency. Investigators soon found that immobilisation was a significant improvement and commercial use of immobilised glucose isomerase made rapid progress. Today several companies produce immobilised enzymes for the isomerisation process. Seidman[14] has made an excellent summary of these early technological developments.

Manufacture

Granular starch suspended in water is continuously liquefied by acid and/or α-amylase treatment at elevated temperatures. This resulting liquefied hydrolysate has a DE in the range 15 to 25. The liquefied product is adjusted for pH and temperature and then saccharified with glucoamylase. The finished saccharified liquor will have a dextrose content of 94 to 96%. It is then refined by filtration plus multiple activated carbon and ion-exchange systems.

The refined dextrose liquor is then pumped through reactors containing immobilised isomerase. The resultant syrup has a fructose content of about 42%. It is further treated with activated carbon and ion-exchange resins before pH adjustment and vacuum concentration to 71% solids. The finished high fructose glucose syrup is shipped directly in bulk to customers, where it is stored at about 30°C.

Analysis

Running parallel with the advances in processing and enzyme technology were equally vital advances in analytical techniques. Particularly the high performance liquid chromatographic separation and quantitative estimation of glucose and fructose. An exceptionally useful analytical system is described by Scobell et al.[19] This system lends itself to micro-processor control and data retrieval. Without such systems the routine control of isomerisation would be cumbersome and expensive.

Physiological Properties

High fructose glucose syrup, like its major components fructose and dextrose, has an energy value of approximately 4 cal/g dry basis. Fructose occurs in honey at high levels in the range 41 to 47% dry basis.[16] It is the predominant sugar in many common fruits, as well as a major sugar in most vegetables. In addition fructose occurs as a moiety of sucrose (i.e. in combination with dextrose).

The absorption of fructose from the small intestine of humans occurs by passive diffusion rather than by active transport, as is the case with glucose.[20] For this reason, at normal concentrations the rate of absorption of fructose is slower than that of glucose.

About 10 to 20% of ingested fructose is metabolised by the human gut wall. The remainder is metabolised almost exclusively by the liver. Fructose is found in minute quantities in the blood since the liver very efficiently phosphorylates fructose to fructose-1-phosphate with the enzyme fructokinase. Liver cells are freely permeable to fructose and probably do not require insulin for initial entry, as is the case with glucose. Some insulin is required for the ultimate utilisation of fructose for energy, the amount of insulin depending upon the extent to which fructose is converted to glucose.[5]

Comparative Sweetness

The sweetness of high fructose glucose syrup easily exceeds that of glucose syrups with no fructose content and also that of dextrose.[9] Organoleptic testing of 42% fructose HFGS, as compared to sucrose, shows that they are equivalent in sweetness at 15% solids and above. At lower solids content, minor differences are sometimes noted by experts between the monosaccharide and disaccharide sweetness. The taste sensation of high fructose glucose syrup is often described as a 'clean' sweetness, i.e. the sweetness is immediate but does not linger.

High fructose glucose syrup has a sweetness value in practice very close to that of sucrose syrups at the same solids content. Comprehensive taste panel evaluation in finished commercial products indicates that the 50/50 combination of high fructose glucose and sucrose syrups is not significantly different from that of just sucrose.[17] Valuable effects also seem to be obtained in formulations involving saccharin and high fructose glucose syrup, e.g. reduced saccharin after-taste plus enhanced total sweetness response.

Storage and Use

The viscosity of high fructose glucose syrup is similar to that of other mono- and disaccharide syrups. The commercial product, containing 42% fructose, has a solids content of 71% and is easily pumped through existing liquid sweetener systems in food plants. Storage at approximately 30°C is the minimum temperature that will prevent crystallisation.

APPLICATIONS OF HIGH FRUCTOSE GLUCOSE SYRUPS

Soft Drinks

Many carbonated and still beverages, representing the complete spectrum of flavours, are made with high fructose glucose syrups as the total sweetener. Typically, some minor reformulation of certain beverages, as compared to a formula using sucrose or medium invert, has been valuable. Other applications indicate a preference for partial replacement in the range 25 to 50%. From its first introduction, high fructose glucose syrup has been used widely in flavoured fountain syrup concentrates of the type that are diluted with carbonated water and consumed like other carbonated beverages.[5, 9]

In still beverages, high fructose glucose syrup is often used with other glucose syrups, dextrose, medium invert or sucrose. It adds a sweetness that complements fruit flavours. Since the composition of high fructose glucose syrup does not change (as compared to the inversion of sucrose) in low pH beverages, the products can be formulated to a sweetness profile that will remain constant. Alcoholic beverages, such as liqueurs, cordials and sweetened wines. may all make use of high fructose glucose syrup.

Ice Cream

Ice cream manufacturers have found high fructose glucose syrup to be a valuable addition to their choice of nutritive sweeteners. It is used along with other glucose syrups in ice cream, ice milk and frozen desserts. Freezing-point depression is minimised by using it with low DE glucose syrup or by retaining some of the sucrose in the formula. The use of high fructose glucose syrup yields ice cream with a smooth, creamy texture and desirable melt-down rates. Dairy manufacturers use it in chocolate milk and other beverages. The amount of chocolate flavour used can be reduced while maintaining taste characteristics.

Pickles and Sauces

Among the first commercial applications of high fructose glucose syrup was its use in sweet pickles and ketchup. High percentages may be incorporated in the manufacture of both products. In sweet pickles, the rapid penetration of the monosaccharides into the cucumber or onion helps reduce processing time and results in a firm, crisp product. Ketchup made with high fructose glucose syrup maintains a deep red colour which gives it a high acceptance rating. Its 'clean' sweetness allows the tomato flavours and added spices to be fully appreciated.

Jams and Preserves

Jams, jellies, preserves and canned fruits are being manufactured with high fructose glucose syrup replacing a major part of the sucrose portion. The enhancement of fruit flavours is important in these applications. Processors find that high quality products are obtained along with ingredient economy.

Second Generation High Fructose Glucose Syrups

Currently, US and Japanese producers are now offering high fructose glucose syrups with fructose levels in the range 55 to 60%, as shown below:

Typical analysis of 55% to 60% HFGS
Solids	77%
Ash (sulphated), DS	0·03%
Fructose, DS	55 to 60%
Dextrose, DS	41 to 36%
Other saccharides, DS	4%

These have been developed for some major applications that have either used high fructose glucose syrup at less than half the total sweetener or have not used it at all. This has occurred primarily with carbonated beverages, where subtle differences particularly in cola formulations are of major concern. The fructose level of these products has been increased while the dextrose level has been decreased. The 55 to 60% products can be shipped at a higher solids level (77%) with little or no tendency to crystallise.

To further serve the food industry 90–95% fructose syrups have been introduced with the following typical analysis:

Typical analysis of 90% + fructose syrup
Solids	75 to 80%
Ash (sulphated), DS	0·03%
Fructose, DS	90 to 95%
Dextrose, DS	7 to 3%
Other saccharides, DS	3 to 2%

This 90% + fructose syrup is essentially non-crystallising. Like other sweeteners, it has an energy value of 4 cal/g dry basis. It is sweeter than any glucose syrup or sucrose or invert sugar syrup and because of its high sweetness is being used in a wide range of speciality foods to reduce calorie levels and yet maintain sweetness. Reduced calorie fruit-flavoured beverages of all types, carbonated, still or alcoholic, are either in commercial production or soon will be. Reports are that flavour and body characteristics are being developed that are new and appealing.

Reduced calorie foods, e.g. jellies, preserves, salad dressing and table syrup, are now being manufactured with 90%+ fructose syrup and are recognised as having quality improvements.

It is generally accepted that the sweetness of fructose is greater in foods at low temperatures. Frozen desserts of various types can, therefore, utilise this characteristic of 90%+ fructose syrup. The developing market for frozen yogurt is an example where this new sweetener has been accepted, providing adequate sweetness with the use of lower quantities of sweetener. This also lowers the calorie content of the frozen yogurt.

These 90%+ fructose syrups will be used by processors who wish to develop new foods with fewer calories and yet maintain a desirable sweetness and by food researchers who need to develop new foods to expand their markets. The combination of its high sweetness, humectancy, flavour enhancement and ingredient compatibility will place it in a favourable position.

THE FUTURE FOR EUROPE

Political Block

Advances in enzyme and process technology have had a major impact on the production and application of sweeteners from starch. The impact of such 'Biotechnology' worldwide is not yet finished. Increased availability of the higher fructose (55%+) syrups will be interesting. Such developments in the USA, Canada, Japan and Eastern Europe are watched by the EEC with a mixture of concern and envy. Production of first generation high fructose glucose syrups was limited severely (and retro-actively) by the European Commission. The existing Common Agricultural Policy guarantees that further generations of high fructose glucose syrups would be still-born.

Finished Imports

Nevertheless these higher fructose glucose syrups will be available in the EEC in the form of imported finished food/drink products. Developing countries supplying the EEC will use the latest biotechnology syrups in, e.g. preserves jam, confectionery, sauces and fruit nectars. It is the EEC food and drink industries that will suffer (just as their counterparts in the automobile industry are suffering) because the final customer sees and buys the advantages of improved technology in imported finished products.

The Way Forward
We stand at the threshold of exciting and far-reaching advances in biotechnology, which are relevant to the food and drink industries and more widely. The EEC governments must be vigorously persuaded to encourage development of such technology leading to commercial use. The advanced starch extraction and syrup producing companies can contribute markedly to such progress and thereby strengthen Europe's future.

REFERENCES

1. KNIGHT, J. W. (1969). *The Starch Industry*, Pergamon Press, Oxford.
2. BANKS, W. and GREENWOOD, C. T. (1975). *Starch and its Components*, Edinburgh University Press.
3. PALMER, T. J. (1970). In: *Glucose Syrups and Related Carbohydrates*, Eds Birch, G. G., Green, L. F. and Coulson, C. B. Applied Science Publishers, London, 23–8.
4. HOWLING, D. (1979). In: *Sugar: Science and Technology*, Eds. Birch, G. G. and Parker, K. J., Applied Science Publishers, London, 263–72.
5. CORN REFINERS ASSOCIATION INC. (1978). Seminar Proceedings: Products of the Corn Refining Industry, Washington DC.
6. NIEMAN, C. (1960), *Manuf. Confectioner*, **40**(8), 19–24, 43–6.
7. PALMER, T. J. (1975). *Process Biochem.*, **10**, 19–20.
8. MAIDEN, A. M. (1971). *The Brewer*, **57**(677), 76–83.
9. PALMER, T. J. (1975), *Soft Drinks Trade Journal*, **29**(10), 358.
10. SMITH, J. B. (1977), *The Brewer*, **63**(748), 50–5.
11. PARSONS, F. M. and FORE, H. (1963), *Lancet*, **2**, 386–7.
12. SHALLENBERGER, R. S. and BIRCH, G. G. (1975). *Sugar Chemistry*, Avi Publishing, Westport, 46–88.
13. NICHOLSON, D. E. (1977). *A Guide to Metabolic Pathways and Coenzymes*, Koch-Light Laboratories, England, 2–4.
14. SEIDMAN, M. (1977). In: *Developments in Food Carbohydrate—1*, Eds Birch, G. G. and Shallenberger, R. S. Applied Science Publishers, London, 19–42.
15. JUNK, W. R. and PANCOAST, H. M. (1973). *Handbook of Sugars*, Avi Publishing, Westport, 157–80.
16. WHITE, J. W. and UNDERWOOD, J. C. (1974). In: *Symposium: Sweeteners*, Ed. Inglett, G. E., Avi Publishing, Connecticut, 120.
17. ROBINSON, J. W. (1975), *Food Engineering*, **47**(5), 57–61.
18. BUCKE, C. (1979). In: *Developments in Sweeteners*, Eds. Hough, C. A. M.,

Parker, K. J. and Vlitos, A. J., Applied Science Publishers, London, 43–68.
19. SCOBELL, H. D., BROBST, K. M. and STEELE, E. M. (1977). *Cereal Chem.*, **54**(4), 905–17.
20. NIKKILA, E. A. and HUTTUNVEN, J. K. (Eds.) (1972). *Symposium: Clinical and Metabolic Aspects of Fructose* Suppl. 542, Acta Medica Scandinavica.

6

Lactose and Lactitol

P. LINKO
*Department of Chemistry,
Helsinki University of Technology,
Espoo 15, Finland*

ABSTRACT

The estimated annual worldwide availability of lactose as a by-product of the dairy industry is several million tons. Whey contains about 4·8% of lactose, which may be purified by crystallisation. Food industry applications, both of pure lactose and lactose-containing dairy by-products, have markedly increased during the last two decades. Lactose is a reducing sugar and thus may be employed to improve colour formation through Maillard reactions. Other food applications are based on its relatively low sweetness and unique crystallisation characteristics. Lactose may contribute body and viscosity to the product without excessive sweetness, and it has been used to improve texture and structure. With its bland flavour, it is an excellent carrier and stabiliser of many aromas and pharmaceutical products. The non-hygroscopic crystalline α-lactose is ideal as an anticaking agent in many dry powders.

The relatively low solubility of lactose limits its use in certain applications. Another limiting factor is the inability of lactose-intolerant people to digest milk sugar. However, recent advances in enzyme engineering have resulted in hydrolysed lactose products without these adverse effects. Lactose is little fermented by bakers' yeast and during brewing, which may be used to advantage. Many fermented foods are based on the ability of other microorganisms to utilise lactose. An interesting recent approach is the utilisation of lactose-containing dairy by-products for biotechnical production of single cell protein, ethanol, and other useful products.

Lactose can be isomerised to lactulose, and hydrogenated to the *polyhydric alcohol, lactitol, both of which have been reported to be poorly absorbed from the digestive tract, and to possess bifidogeneous activity when fed to young infants. Lactitol does not appear to affect blood glucose or insulin levels, and it has shown similar anticariogeneity to that of xylitol.*

INTRODUCTION

The lactose (milk sugar) content of milk from different mammals varies from 0 to about 9%. Fluid whole cow's milk contains about 4·9% of lactose, non-fat milk solids contain about 53%, and dried whey as much as 73%. Human milk contains about 6·7% of lactose, considerably more than cow's milk, whereas the milk of sea lions and some other mammals has no lactose at all.[1] Small quantities of lactose have been detected in blood and urine, but other sources of lactose are rare.[1] The potential annual worldwide availability of lactose as a dairy industry by-product is of the order of five million tons, of which only about 100 000 tons were produced in 1973 in the EEC countries,[2] and roughly an equal quantity in 1970 in the USA.[3] The principal commercial source of lactose is cheese whey, from which it may be obtained either by direct crystallisation or, after the removal of protein, either by ultrafiltration or by precipitation. Considerable amounts of lactose are also utilised in the form of dried milk and whey products, and, owing to its high lactose content, whey powder is often the choice as the most economic source of lactose to improve body, texture, viscosity or mouth feel of a variety of foods. The recent increased interest in alternative carbohydrate-based sweeteners has paralleled the needs to improve whey utilisation, and the demand for lactose is likely to grow in the future.

Lactose was discovered in milk in 1619 by Bartoletti,[4] and identified as a sugar in 1780 by Scheele. Lactose is hydrolysed to glucose and galactose, isomerised in alkaline solution to lactulose, and catalytically hydrogenated to the corresponding polyhydric alcohol, lactitol (Fig. 1). A complete coverage of the vast literature available on lactose and its derivatives is beyond the scope of this paper. Nickerson[1] has recently written an excellent review on lactose, β-galactosidase technology has been extensively treated by Shukla[5] and lactose hydrolysates by Harju and Kreula.[6] Lactulose

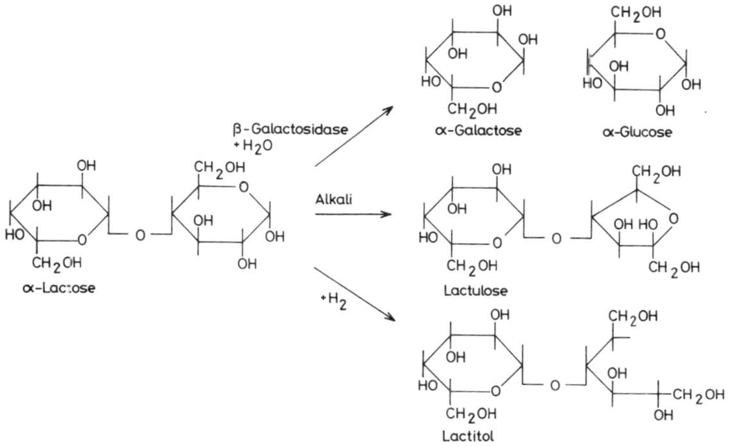

FIG. 1. Lactose transformations.

has been studied by Adachi and Patton,[7] and lactitol by Linko et al.[8] This paper describes the current state of the art of lactose utilisation, emphasising food applications.

LACTOSE

Characterisation of Lactose

Lactose (O-β-D-galactopyranosyl-[1→4]-D-glucopyranose), $C_{12}H_{22}O_{11} \cdot H_2O$ (MW 342·3) is a reducing disaccharide, normally crystallising as the monohydrate, that yields glucose and galactose on hydrolysis. The knowledge of the physical properties of lactose is invaluable for the understanding of its behaviour in food systems. Lactose exists both in α- ($[\alpha]_D = +89\cdot4°$) and β- ($[\alpha]_D = +35°$) anomeric forms. The stable α-lactose monohydrate (m.p. 201·6°C) crystallises at $< 93\cdot5°C^{1,9}$ as water-soluble, hard, white, rhombic crystals of mild sweet taste. Anhydrous α-lactose is more readily soluble in water, very hygroscopic, and melts at 222·8°C. β-Lactose anhydride (m.p. 252·2°C) is obtained on crystallisation at $> 93\cdot5°C$. An equilibrium state ($[\alpha]_D = +55\cdot5°$) exists with 37·75% of α-lactose in solution at 25°C. The conversion is very rapid at 70°C. Rapid drum- or spray-drying of a lactose solution in the equilibrium state results in a hygroscopic amorphous lactose glass

(Fig. 2), in contrast to the crystalline non-hygroscopic form obtained by slow drying. The physical state of lactose thus has a marked effect on the sorption properties, and the unique behaviour is also of significance in drying and in the stability of milk products.[10]

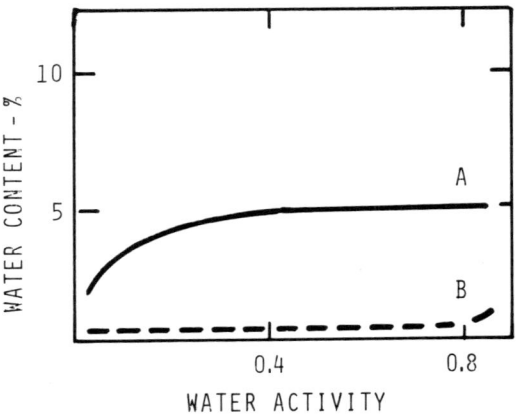

FIG. 2. Adsorption isotherms of (A) amorphous and (B) crystalline lactose (25°C).

The anhydrous unstable form of α-lactose, obtained by heating *in vacuo*, may be recrystallised from dry methanol to give the stable anhydride.[11] Nickerson[1,12], and Thurlby[13] have reviewed lactose crystallisation in detail. Characteristically, lactose has the ability to form supersaturated solutions.[12] The work of Bouldoires[14] in which crystallisation in pastes was followed by dielectric measurements has recently been applied to improve the chocolate manufacturing process.[15] Thermal analysis both by thermogravimetric analysis[16] and differential scanning calorimetry[17] has been applied to determine the crystalline lactose content in whey powder. This technique has also been employed to show the presence of the β-anomer in α-lactose hydrate.[18] Furthermore, an increase in the β-form was observed to result in an increased formation both of lactulose and epilactose (*O*-β-D-galacto-pyranosyl-[1→4]-D-mannose) during thermal treatment. The relative quantities of α- and β-lactose have also been determined by gas–liquid chromatography.[19]

Solubility

The solubility behaviour of lactose differs considerably from that of sucrose. Lactose is significantly less soluble, but the relative change in solubility with increasing temperature is higher (Fig. 3).[20] To obtain a saturated lactose solution at a given temperature takes about three times as long as it does with sucrose. The markedly lower solubility at room temperature may be used to advantage in some icings, and in coating sugar. In other icings, the crystallisation of lactose on cooling may result in grittiness. Lactose crystals of $\phi > 10\,\mu m$ may also be the cause of so-called sandiness in ice-creams, and may facilitate protein coagulation on reconstitution of frozen milk concentrates.[5]

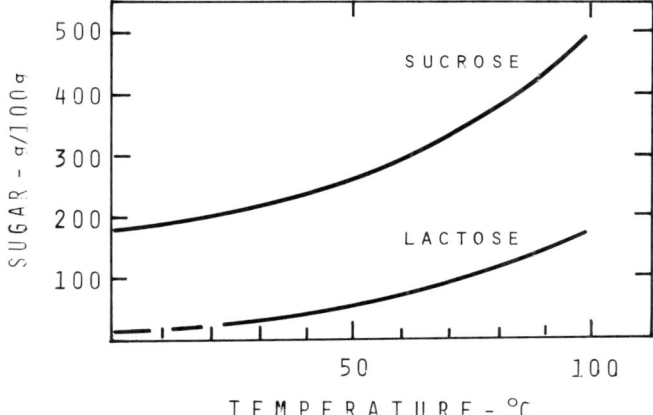

FIG. 3. Effect of temperature on the solubility of lactose and sucrose in water.

Sorption properties

Lactose adsorbs more water at 60% RH than sucrose, but at 100% RH both glucose and sucrose are able to hold more water than lactose (Fig. 4).[20] The differences in the water adsorption behaviour of initially amorphous lactose anhydride and crystalline α-lactose monohydrate are clearly seen in Fig. 5.[21]

Sugars are well known as excellent carriers of various flavours. Different sugars vary considerably, however, in their ability to adsorb aroma compounds.[2,22] Lactose is generally regarded as a better carrier than fructose, glucose or sucrose.[22,23] On the other

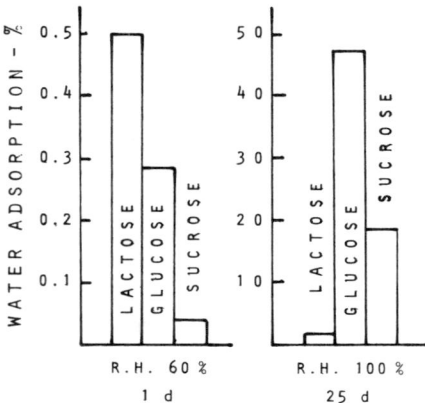

FIG. 4. Water adsorption capacity of lactose, glucose, and sucrose at 20°C.[20]

hand, data on the flavour-enhancing characteristics of lactose are somewhat contradictory.[20] Ash[20] also pointed out that food grade lactose frequently has a typical whey flavour that is not, however, normally carried over to the product at proper dosages.

Relative sweetness

The relatively low sweetness of lactose may be an advantage or a disadvantage, depending on the application. At 1% concentrations, the relative sweetness value frequently given to lactose is about 0·15 (sucrose = 1), but it is important to realise that sweetness is also a function of concentration (Fig. 6),[20] temperature and pH, and that the relative sweetness of lactose may thus, in practice, vary considerably (Fig. 7). According to the author's own results, the relative sweetness of lactose at 10% (w/v) level at 23°C was only about 0·1,[8] but it has been reported to approach a value of 0·3 at 40% concentration.[20] Shah and Nickerson[24] observed a synergistic effect on the sweetness of ice cream with a mixture of galactose, glucose, and lactose, while Harrison[25] and Pangborn[26] found no synergism between lactose and galactose, or lactose and xylitol.

Nutritional and physiological considerations

Although lactose is digested more slowly than sucrose, the reported caloric value (1·566 kJ/100 g) is nearly the same as for sucrose (1·650 kJ/100 g). Consequently, lactose is as good an

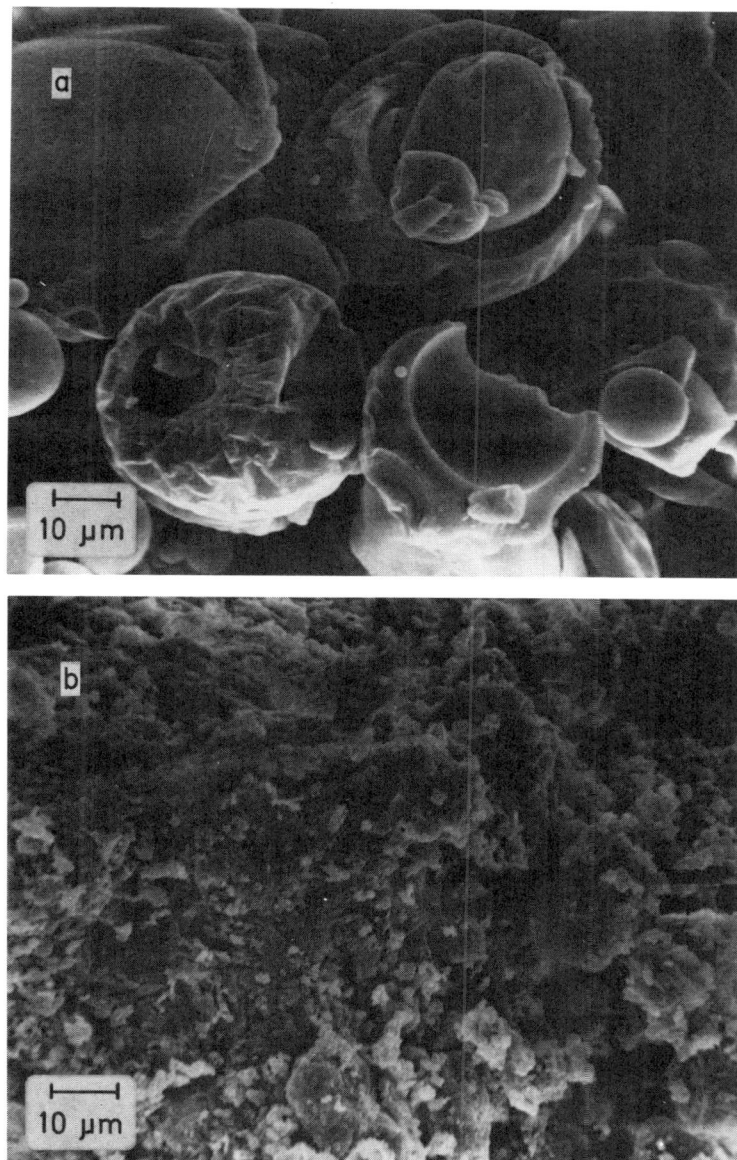

FIG. 5. Scanning electron micrographs of (a) dry and (b) wetted (a_w 0·5) amorphous lactose.

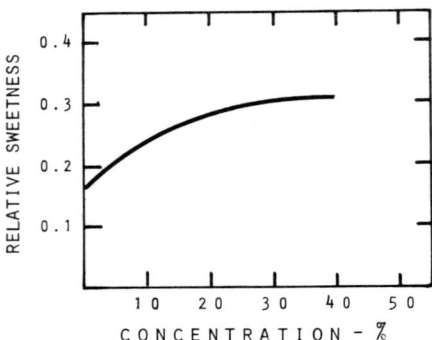

FIG. 6. The effect of concentration on relative sweetness of lactose at 20°C (sucrose = 1·0).

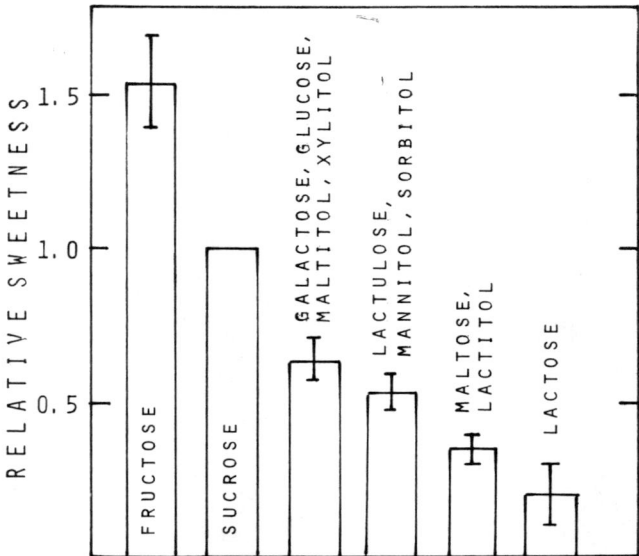

FIG. 7. Relative sweetness of various sugars and sugar alcohols.

energy source as other sugars, and cannot be considered as a low-calorie sweetener. Because of the relatively slow absorption of lactose, comparable to that of starch, lactose has been recommended for some diabetic food applications.[27,28]

It was suggested as early as 1913 that lactose would have a specific role in facilitating the absorption of calcium and essential trace metals from the intestine.[29,30] Apparently the action is non-specific, inasmuch as lactose has also been shown to increase the intestinal absorption of certain toxic elements such as lead.[31]

One litre of typical cow's milk contains about 40 g of lactose, which thus is an important carbohydrate in the diet of children and many adults. Nevertheless, lactose intolerance due to the inability to utilise lactose because of insufficient jejunum β-galactosidase is widespread.[5,32-35] Undigested lactose is passed into the blood, urine, and the large intestine, where it is responsible for osmotic dehydration, poor calcium absorption owing to resulting low acidity, and fermentative diarrhoea characterised by flatulence, belching and cramps. Consequently, Pyke[36] considers lactose poisonous to severely lactose-intolerant individuals. However, lactase activity normally reaches a minimum at about 2 to 4 years of age, and genuine lactose intolerance should be distinguished from the one acquired as a consequence of a low lactose diet.[22,37] Untreated, β-galactosidase deficiency in infancy may be rapidly fatal owing to dehydration and loss of energy.

Galactose is a necessary constituent of connective tissue mucopolysaccharides, and of the cerebrosides of brain and nerve tissue. Glucose and galactose are believed to compete for the same active transport system across the gut wall, and an increase in glucose intake suppresses blood galactose level. Galactose is primarily metabolised in liver and, unlike glucose, does not appear to initiate the release of insulin. However, the blood lactic acid level increases after galactose intake, suggesting that glucose would be a better sugar for exercise. A rare congenital galactosaemia is a hereditary disease due to a single recessive genetic defect. The lack of galactose-1-phosphate-uridyl-transferase activity results in the accumulation of galactose-1-phosphate with excessive galactose intake and, subsequently, in liver and brain damage and cataracts.

Lactose is a mild laxative and can be utilised by intestinal bacteria to form lactic acid to maintain the low pH necessary to prevent putrefaction.[37] In addition it encourages peristalsis and thus prevents constipation. It is also believed that the formation of dental plaque is not enhanced by lactose.[38] The current nutritional status of lactose and galactose has been discussed by Macdonald.[39]

Manufacture of Lactose

Modern lactose production technology is based on crystallisation from cheese whey either directly or after the removal of proteins.[1,40-46] Purification by precipitation, ultrafiltration, ion exchange and electrodialysis may precede whey concentration, or whey proteins may be kept in solution during crystallisation by adding polyphosphates.[47,48] After ultrafiltration, calcium phosphate may be removed by ion-exchange or electrodialysis, and reverse osmosis may be employed for simultaneous preconcentration.[43,49] Extraction of lactose and crystallisation from organic solvents has also been employed. A crystal size above 100 μm facilitates separation from the mother liquor. Optimisation of the cooling of the concentrate is important in controlling crystallisation of α-lactose. At low temperatures the rate of mutarotation determines the rate of α-lactose crystallisation.[42] Typical yields are of the order of 70 to 80%, with a significant improvement if the whey is deproteinised by ultrafiltration prior to crystallisation.[41,43] Continuous processes have also been developed.[42,45]

Most lactose on the market is the stable α-monohydrate, but β-lactose is also produced for special applications. β-Lactose is obtained in stable form on crystallisation from solutions at above 93·5°C.[1,50]

Utilisation of Lactose

Lactose has been used for medical purposes for many years.[51] Non-hygroscopic crystalline α-lactose monohydrate is an excellent anticaking agent,[51] and it has been used as a carrier for antibiotics.[52] Lactose has excellent tablet-forming characteristics which may be modified by the crystalline composition.[53] One of the most important applications of lactose today is the use in infant formulas to simulate the composition of human milk.

Because it has no characteristic flavour of its own, lactose is compatible with many food flavours and other additives. It is an ideal carrier for artificial sweeteners to give free flowing powders.[51] Flavourings protected through agglomeration and instantising by co-spray drying with concentrated lactose solution may be easily dusted directly on to hot and moist foods. The hygroscopicity of amorphous lactose glass in spray-dried milk or lactose solutions can also be utilised to advantage: careful slow addition of moisture to particles agitated in an air-stream produces porous aggregates that on redrying yield 'instant' milk powder.[55] Furthermore, lactose may be added to other food solutions prior to drying for instantising.[51]

In baked goods, lactose can contribute to flavour, texture, appearance, shelf-life, and toasting properties.[20, 56] Larsen et al.[57] observed as early as 1949 that milk, even when heated, slightly depressed loaf volume in straight dough processes. Guy et al.[56] observed similar behaviour in sponge dough baking, and recommended decreased whey lactose levels in certain baking applications. Nevertheless, Holmes and Zahringer[58] were able to show by response surface methodology that an increase in loaf volume may be realised with lactose by optimising the combined effects of lactose, sucrose, and shortening. Furthermore, owing to its unique water-binding properties, lactose appears to improve tenderness and prolong the shelf-life of many products, and also seems to permit greater tolerances to variations in other ingredients. Lactose in non-fat dry milk has been found to increase both the tenderness and volume of biscuits and donuts.[59, 60]

Although a number of microorganisms are able to hydrolyse lactose, lactose is not fermented by bakers' yeast. Consequently, as a reducing sugar lactose is available for crust browning without resulting in excessive sweetness of the product. According to Hugunin,[61] lactose alone produces a uniform golden brown crust of pleasant flavour. A more pronounced browning effect is obtained in the presence of whey proteins. Lactose also improves toasting quality owing to the Maillard reaction. Up to 50% of sucrose in cake batter could be replaced with lactose, with a decrease in sweetness noticeable only in unfilled cakes; 25 to 30% replacement gave excellent results for small medium-shortening cakes with a sweet topping.[20] Lactose in chocolate cake batters, however, resulted in a light crumb colour.

Lactose may also be used to improve browning in other food systems. Jelen and Jadhar[62] reported that the addition of diluted cottage cheese whey, or a mixture of lactose and glutamic acid, improved the colour of french-fried potatoes. Beneficial results with lactose have also been observed in microwave cooking in general.[51] The addition of about 10% of lactose to raw fish has been claimed to improve flavour,[63] and Jelen and Breene[64] observed that the addition of lactose improved the texture of dill-pickle without imparting excessive sweetness. Lactose adds mouth feel and viscosity to a number of foods and can be successfully used in custard, fruit pies, jams, and preserves,[29,65] and in certain icings.[66] In addition to its mild sweetness and to its effect on consistency, lactose also enhances fruity flavours,[67] but its low solubility may result in crystallisation on long standing at high replacement levels. Consumer acceptability of milk and fermented milk based beverages has been attributed to lactose,[66,68,69] and lactose-containing chocolate drinks have been judged superior to those with no lactose.[70] The ability of lactose to stabilise proteins has also been utilised in the preparation of certain foams and whipped dessert products.[71,72]

LACTOSE HYDROLYSATES

Lactose intolerance and certain technological problems related to the low sweetness and solubility of lactose have prompted the development of processes for the production of lactose hydrolysates suitable for a variety of applications.[5,6,73–75]

Hydrolysis of Lactose

Lactose may be hydrolysed to galactose and glucose by hydrogen ion catalysis[6,44,76–78] or enzymatically with β-galactosidase.[5,6,77,79] Acid hydrolysis generally requires high temperatures or a long reaction time with diluted acid, and only relatively pure lactose may be used. Continuous processing employing strong cation-exchange resins at 90–100°C with a residence time of 0·5 to 2·0 h has been developed.[80,81] The recent advances in immobilised β-galactosidase technology have made continuous enzymatic processing both of lactose and lactose-containing dairy products possible.[5,6,82–86] The use of immobilised β-galactosidase to hydrolyse lactose in dairy products in order to increase digestibility, solubility, and sweetness, has marked technological and economic potential. Today, Centrale del Latte in Milan, Italy, treats milk semi-continuously in a plant based on recirculation reactors of cellulose triacetate fibre carrying entrapped *Kluyveromyces lactis* yeast β-galactosidase.[87] The Valio Cooperative Dairies Association in Finland operates a plant for the continuous hydrolysis of acid whey lactose with phenol formaldehyde resin immobilised *Aspergillus niger* β-galactosidase column reactors, using a process developed in co-operation with the author's laboratory and the Technical Research Centre of Finland.[6] Two semi-industrial plants based on *A. niger* lactase covalently bound to controlled pore size silica are in operation in Europe.[85]

A detailed investigation into the kinetics of milk lactose hydrolysis with soluble *K. lactis* β-galactosidase has been published,[79] and a method to immobilise β-galactosidase active *K. fragilis* yeast cells within cellulose di- or triacetate beads, as a relatively stable active preparation, developed (Fig. 8).[86]

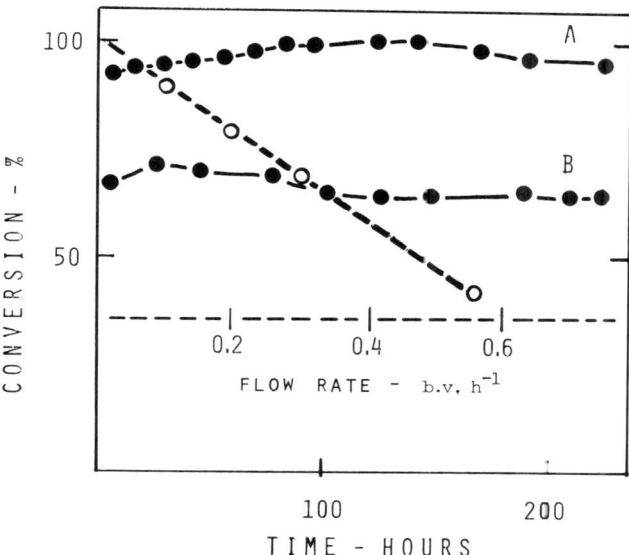

FIG. 8. Examples of the continuous hydrolysis of lactose with cellulose diacetate (●, A) and triacetate (●, B) bead immobilised *Kluyveromyces fragilis* cells, and the degree of conversion as a function of flow rate (○) (20°C pH 6·5).

Properties of Lactose Hydrolysates

In a typical process lactose is hydrolysed to about 80% conversion. Optimum solubility characteristics are reached at about 75% hydrolysis.[77,88] At lower conversions, lactose crystallises first, and, at a higher degree of hydrolysis, galactose crystallises first. Only within a relatively narrow range of about 55 to 65%

solids, is microbiologically stable syrup that does not easily crystallise obtained. It has also been observed that about 80% hydrolysis is sufficient for most lactose-intolerant individuals. Owing to synergism the sweetness of the syrup is higher than expected, and at 50% total solids it approaches that of sucrose.[77] An increase in conversion above 80% has little effect on sweetness, while the production costs increase sharply.[6] If a higher degree of sweetness is desired, however, glucose may be enzymatically isomerised to fructose.[89,90] In the author's experiments, the relative sweetness of lactose hydrolysate of about 85% conversion (10% solids, 25°C, pH 5·5) was 0·60–0·65 (sucrose 1·0), and after 40% of the glucose was isomerised, 0·7–0·8. If the glucose content of the syrup was increased to 54% of total solids, the relative sweetness after isomerisation approached that of sucrose.

In confectionery applications the humectant properties of hydrolysed whey lactose syrup were equal to or better than those of invert sugar.[77,91,92] Maillard browning is enhanced by lactose hydrolysis,[93] and the viscosity of concentrated solutions is significantly decreased.[91]

Uses of Lactose Hydrolysates

At least 20% of the sucrose in ice cream can be replaced by hydrolysed lactose.[6] If lactose is partially hydrolysed during yogurt manufacture, the fermentation time is shortened,[94,95] and the taste of the final product improved.[94] Unlike lactose, both glucose and galactose are fermented by bakers', brewers' and wine yeasts, and hydrolysed lactose syrups have been employed as sugar adjuncts in beer[96,97] and wine[98,99] fermentations, and in many bakery products.[100,101] In beer, up to 50% of normal wort may be replaced by hydrolysed lactose syrup.[96,97] Other suggested food applications include soft drinks, confectionery, and fruit preserves.[73] Furthermore, inasmuch as many animals are unable to digest large quantities of lactose, the whey ratio in their daily intake may be significantly increased if spray-dried whey powder in the ration is replaced by hydrolysed whey syrup.[6]

LACTULOSE

Lactulose has not been found in nature, but it has been reported to be formed by isomerisation of lactose during thermal processing and long storage of milk and certain other dairy products.[7,102,103] A lactulose content of up to 8% of the total carbohydrates has been found in infant formulas.[103]

Characterisation of Lactulose

Lactulose (O-β-D-galactopyranosyl-[1→4]-D-fructofuranose) is water-soluble, highly hygroscopic, and difficult to crystallise from water solutions. Colourless, hexagonal crystals of α-lactulose melting at 158°–165°C have been obtained from 50% methanol.[104] The value of the specific rotation, $[\alpha]_D^{22} = -23 \cdot 9°$, on dissolution changes, owing to mutarotation, to the equilibrium value of $[\alpha]_D^{22} = -51 \cdot 5°$ on standing. The relative sweetness of lactulose in the concentration range of 5 to 35% (w/v) is about equal to that of sorbitol.[105]

Ross et al.[106] reported that lactulose would be an excellent humectant, giving an $a_w = 0 \cdot 85$ at half the corresponding sucrose concentration. More recently, Chirife and Ferro Fontan[107] showed that, although lactulose exhibits a negative deviation from Raoult's law slightly higher than that of sucrose, the effect on water activity is far less dramatic than that claimed by Ross et al.[106] Lactulose is hydrolysed by *Escherichia coli* β-galactosidase to yield galactose and fructose. The formation of fructose may be utilised in the enzymatic determination of lactulose in the presence of lactose.[108]

Manufacture of Lactulose

Lactulose is currently produced by alkaline isomerisation of lactose.[109,110] A method has also been described, in which the remaining lactose is oxidised enzymatically to lactobionic acid, which is then removed by ion-exchange or by precipitation as the calcium salt.[111] The alkaline isomerisation reaction is slow, and typical yields are less than 25%. However, yields of up to 80% in 3 h have been reported by treating a 20 to 30% lactose solution at 50–100°C, pH 11·5, in the presence of sodium or calcium aluminates.[112] Vaheri and Kauppinen[113] have recently shown that lactose may also be enzymatically converted to lactulose by utilising the tranferase activity of β-galactosidase and using fructose as the acceptor.

Physiological Effects and Uses of Lactulose

The most important current applications of lactulose are in

infant formulas,[108,113] and in the treatment of portal systemic encephalopathy (PSE)[114] and constipation.[115,116] Lactulose is not hydrolysed in the small intestine, but it is metabolised by the colon bacteria to acetic and lactic acids, a process associated with a decrease in pH, an increase in osmotic pressure, and changes in faecal microflora.[117,118] The beneficial effects of lactulose in the diet of young children are believed to be due to the formation of pure *Lactobacillus bifidus* flora in the intestine.[119,120] Since the introduction of lactulose in 1957 in clinical trials for infants,[121] lactulose has been used in infant formulas.[108,122] The clinical uses have been reviewed by Kardel.[122]

LACTITOL

The polyhydric alcohols maltitol, mannitol, sorbitol and xylitol are industrially produced from the corresponding sugars by catalytic hydrogenation, and find a number of food and pharmaceutical applications.[123–125] Recently, hydrogenated syrups tailor-made for varying sweetness and functional properties have been developed.[124,126,127] Owing to the abundant availability of lactose, lactitol offers an interesting alternative for a variety of applications.[128–133]

Characterisation of Lactitol

Lactitol (*O*-β-D-galactopyranosyl-[1→4]-D-glucitol), $C_{12}H_{24}O_{11} \cdot H_2O$, (MW 344, melting range 94–97°C, $[\alpha]_D^{20} = +16 \cdot 0°$) forms colourless, odourless, non-hygroscopic crystals. It is slightly less sweet than sorbitol (Fig. 7) and more stable than lactose. Lacking a free carbonyl group, it does not undergo the Maillard reaction. Heating lactitol in dilute solution at 100°C, pH 13,[132] and at 50°C in 3 M hydrochloric acid or in citrate buffer of pH 4·8[133] resulted in no discolouration. Pure lactitol heated at 150°C undergoes gradual colour formation typical of other sugar alcohols. It decomposes at temperatures above 200°C.

Lactitol is readily soluble in water and freely miscible with other sugar alcohols and sugars. Its solutions are slightly less viscous than those of lactose of equivalent concentration, but significantly more viscous than sorbitol solutions.[132,133] Anhydrous lactitol is hygroscopic, and at low relative humidities it adsorbs more water than xylitol. However, at 97% RH the humectant properties of lactitol and xylitol are very similar.

Lactitol yields D-galactose and D-sorbitol on hydrolysis, both widely occurring in nature. Karrer and Büchi[134] were first to report that lactitol would be only slowly hydrolysed by β-galactosidases. According to Shukla,[5] β-galactosidases of different origin vary in their response to different aglucone moieties in the substrate, and Saijonmaa et al.[131] showed that different lactases vary considerably in their ability to hydrolyse lactitol. *Aspergillus niger* lactase LP (Rapidase) hydrolysed lactitol at nearly the same rate as lactose, whereas both *Kluyveromyces lactis* yeast β-galactosidase Maxilact 40 000 (Gist-Brocades) and bacterial *E. coli* lactase (Boehringer Mannheim GmbH) hydrolysed lactitol only slowly. Accordingly, *A. niger* was able to utilise both lactose and lactitol to about the same degree,[8] whereas such typical lactose-fermenting yeasts as *K. fragilis* and *Candida kefyr* metabolised lactitol slowly. It is, however, of interest to note that *E. coli* is able to adapt to utilise lactitol.[8]

Preparation of Lactitol

Lactitol is obtained by the reduction of the glucose moiety of lactose. In 1872 Bouchardat[135] was the first to reduce lactose with sodium amalgam. Later developments employing electrolytic and sodium borohydride reduction, and catalytic hydrogenation have recently been reviewed by Linko et al.[8] The catalytic hydrogenation of 30% (w/v) lactose solution with Raney nickel catalyst was shown to be optimal, yielding about 97% of lactitol in 6 h, at 100°C, under a pressure of 8825 kPa.[8,131] At higher pressures and temperatures the lactitol yield was sharply reduced owing to epimerisation to lactulose and hydrolysis to galactose and glucose, all of which were partially hydrogenated to the corresponding polyhydric alcohols. Above about 120°C under a pressure of 8825 kPa, lactitol formation was negligible (Table 1).

Potential Applications

A variety of food formulations with lactitol, based on its mild, pleasant sweetness and excellent thermal stability, have been presented.[128,136] Instant soups and beverages, ice cream and sherbets, dark chocolates, and many bakery products are among applications cited. Lactitol also retards crystallisation and

TABLE 1
EFFECT OF TEMPERATURE ON CATALYTIC
HYDROGENATION OF LACTOSE TO LACTITOL
(RANEY Ni, LACTOSE 300 g dm^{-3}, 8825 kPa, 6 h)

Temperature (°C)	Lactose (%)	Lactitol (%)
50	80	20
80	44	56
100	3	97
120	2	3
140	1	2

improves moisture retention, and in many cases sucrose may be substituted with lactitol with little if any modifications in the recipe. According to Velthuijsen,[132] lactitol enhances chocolate flavour and extends shelf-life when substituted at 50% level of total sugar, with no process alterations required. In the manufacture of chewing gum and hard candy, glucose and sucrose could be completely replaced by lactitol, and synthetic sweeteners could be used for additional sweetness, if desired. The replacement of half of the sugar in strawberry jam with lactitol resulted in a pleasant-tasting semi-sweet product. Lactitol has also been successfully tested in light biscuit-like products for diabetics.

Lactitol has been reported to be the only disaccharide-derived polyol that is not fermented by *Streptococcus salivarius*, *S. sanguis*, *S. mitis*, *Lactobacillus fermenti*, and *L. avei*, nor by saliva,[137] and the apparent low cariogenicity has been confirmed by clinical trials.[132] It is of special interest that lactitol has also been shown to exhibit similar bifidogeneous activity as lactulose,[138] since pure lactitol may be more easily obtained in high yields than lactulose. Lactitol, like lactose, can also be used in the synthesis of surface-active higher fatty acid esters both for food emulsifiers and detergents.[132,139,140]

CONCLUSIONS

It has been shown that lactose and its derivatives offer a wide variety of possibilities for the utilisation of dairy industry by-products. The recent advances in β-galactosidase technology in particular have resulted in entirely novel approaches. The corresponding polyhydric alcohol, lactitol, also merits further investigation for its promising potential uses both in food applications and as a raw material for emulsifiers.

ACKNOWLEDGEMENT

The author is grateful to Valio Laboratory, Finland for cooperation and encouragement, without which this work would not have been possible.

REFERENCES

1. NICKERSON, T. A. (1978). In: *Fundamentals of Dairy Chemistry*, Eds Webb, B. H., Johnson, A. H. and Alford, J. A., Avi Publ. Company, Westport, Conn., 273–324.
2. GOLLER, H.-J. and KUBE, J. (1974). *Ind. Obst u. Gemüseverwertung*, **59**, 650.
3. BATZ, W. F. (1970). *How Americans use Their Dairy Foods*, National Dairy Council, Chicago, Ill.
4. ANON. (1957). *Ullmanns Encyklopädie der technischen Chemie*, Vol. 9, Erban & Schwarzenberg, Munich, 665.
5. SHUKLA, T. P. (1975). *CRC Crit. Rev. Fd Technol.*, **5**, 325.
6. HARJU, M. and KREULA, M. (1980). In: *Carbohydrate Sweeteners in Foods and Nutrition*, Eds Koivistoinen, P. and Hyvönen, L., Academic Press, New York, 233–42.
7. ADACHI, S. and PATTON, S. (1961). *J. Dairy Sci.*, **44**, 1375.
8. LINKO, P., SAIJONMAA, T., HEIKONEN, M. and KREULA, M. (1980). In: *Carbohydrate Sweeteners in Foods and Nutrition*, Eds Koivistoinen, P. and Hyvönen, L., Academic Press, New York, 243–57.
9. SUPPLEE G. C. (1926). *J. Dairy Sci.*, **9**, 50.
10. SALTMARCH, M. and LABUZA, T. P. (1980). *J. Fd Sci.*, **45**, 1231.
11. NICKERSON, T. A. (1978). *XX Intn. Dairy Congr.*, Paris, 76ST.
12. NICKERSON, T. A. and PATEL, K. N. (1972). *J. Fd Sci*, **37**, 693.
13. THURLBY, J. A. (1976). *J. Fd Sci.*, **41**, 38.
14. BOULDOIRES, J.-P. (1980). In: *Food Process Engineering*, Vol. 1, Eds Linko, P., Mälkki, Y., Olkku, J. and Larinkari, J., Applied Science Publishers, London, 307–13.
15. ROSTAGNO, W., CHEVALLEY, J. and BESSON, J. P. (1977). Swiss P. 598 771.
16. ANDERSON, R. A. and BEVIN, E. (1974). *J. Dairy Sci.*, **57**, 786.
17. ROSS, K. D. (1978). *J. Dairy Sci.*, **61**, 255.

18. FERNANDEZ-MARTIN, F., MORAIS, T. and OLAMO, A. (1980). In: *Food Process Engineering*, Vol. 1, Eds Linko, P., Mälkki, Y., Olkku, J. and Larinkari, J., Applied Science Publishers, London, 523–9.
19. BUMA, T. J. and VAN DER VEEN, H. K. C. (1974). *Netherl. Milk Dairy J.*, **28**, 175.
20. ASH, J. (1976). *Food Product Dev.*, **10**(6), 85.
21. POLLARI, T. (1980). *Water Activity of Dried Milk Products* (in Finnish), M.Sc. Thesis, Helsinki Univ. of Technol., Otaniemi, Finland.
22. KUBE, J. and PRITZWALD-STEGMANN, B. (1976). *CCB Rev. Chocolate-confectionery and Bakery*, **1**(4), 30.
23. MAIER, H. G. and RASMUSSEN, H. (1974). *Lebensmittelchemie u. Gerichtliche Chemie*, **23**, 85.
24. SHAH, N. D. and NICKERSON, T. A. (1978). *J. Fd Sci.*, **43**, 1575.
25. HARRISON, S. K. (1978). *The Time-Intensity Characteristics Saccharin, Xylitol, and Galactose and their Effect on the Sweetness of Lactose*, M. Sc. Thesis, Univ. of California, Davis.
26. PANGBORN, R. M. (1979). In: *Sugar: Science and Technology*, Eds. Birch, G. G. and Parker, K. J., Applied Science Publishers, London, 383–99.
27. VON FUNCKE, H. (1973). *Milchwisseschaft*, **28**, 675.
28. GOLLER, H. (1974). *Zucker u. Süsswarenwirtschaft*, **27**, 336.
29. ALI, R. and EVANS, J. L. (1973). *J. Agr. Univ. P. R.*, **57**, 149.
30. ARMBRECHT, H. J. and WASSERMANN, R. H. (1976). *J. Nutr.*, **106**, 1265.
31. BUSHNELL, P. J. and DELUCA, H. F. (1981). *Science*, **211**(1), 61.
32. HUANG, S. S. and BAYLESS, T. M. (1967). *Acta Paediatr. Scand.*, **56**, 488.
33. ROSENSWEIG, N. S. (1969). *J. Dairy Sci.*, **52**, 585.
34. DAHLQUIST, A. and LINQUIST, B. C. (1971). *Acta Paediatr. Scand.*, **60**, 488.
35. BAYLESS, T. M. (1971). *Gastroenterology*, **60**, 65.
36. PYKE, M. (1972). *Poisonous Foods, Nutrition and Food Science*, J. Murray, England.
37. KRETSCHMER, N. (1971). *Gastroenterology*, **61**, 805.
38. KUBE, J. and PRITZWALD-STEGMANN, B. (1976). *CCB Rev. Chocolate-confectionery and Bakery*, **1**(3), 34.
39. MACDONALD, I. (1978). *J. Soc. Dairy Technol.*, **31**(4), 196.
40. WEISBERG, S. M. (1954). *J. Dairy Sci.*, **37**, 1106.
41. NICOLAISEN, B. N. (1975). *North Eur. Dairy J.*, **41**, 125.
42. THURLBY, J. A. and SITNAI, O. (1976). *J. Fd Sci.*, **41**, 43.
43. LANDRÉ, R. (1975). *North Eur. Dairy J.*, **41**, 116.
44. MACBEAN, R. D. (1979). *New Zealand J. Dairy Sci. Technol.*, **14**, 113.
45. MÜLLER, L. I. (1979). *New Zealand J. Dairy Sci. Technol.*, **14**, 119.
46. BRINKMAN, G. E. (1976). *J. Soc. Dairy Technol.*, **29**(2), 101.
47. ZBORALSKI, U. (1957). German P. 1 008 220.
48. STEIN, W. (1962). German P. 1 097 380.
49. NICKERSON, T. A. (1979). *J. Agr. Fd Chem.*, **27**, 672.
50. BELL, R. W. (1930). *Ind. Eng. Chem.*, **22**, 51.

51. NICKERSON, T. A. (1978). *Fd Technol.*, **32**, 40.
52. LEE, D. E. and LILLIBRIDGE, C. B. (1976). *Am. J. Clin. Nutr.*, **28**, 428.
53. FELL, J. T. (1973). *J. Pharm. Pharmac.*, **25**, 109P.
54. WELCH, D. (1965). *Fd Technol. Australia*, **17**, 318.
55. PEEBLES, D. D. (1956). *Fd Technol.*, **10**, 64.
56. GUY, E. J., VETTEL, R. E. and PALLANSCH, M. J. (1971). *Baker's Digest*, **45**(3), 40.
57. LARSEN, R. A., JENNESS, R. and GEDDES, W. F. (1949). *Cereal Chem.*, **26**, 189.
58. HOLMES, D. G. and LOPEZ, J. (1977). *Baker's Digest*, **51**(1), 21.
59. POTTER, G. L. and ZAHRINGER, M. V. (1965). *Cereal Sci. Today*, **10**, 215.
60. HOFFSTRAND, J. T., ZAHRINGER, M. V. and HIBBS, R. A. (1965). *Cereal Sci. Today*, **10**, 212.
61. HUGUNIN, A. G. (1980). *Baker's Digest*, **54**(4), 8.
62. JELEN, P. and JADHAR, S. J. (1974). *J. Fd Sci.*, **39**, 1269.
63. UENO PHARMACEUTICAL CO., LTD (1977). Japan. P. 5 105 461.
64. JELEN, P. and BREENE, W. M. (1973). *J. Fd Sci.*, **38**, 99.
65. RANDERIA, B. V. (1966). *Confectionery Prod.*, **32**, 845.
66. REGER, J. V. (1958). *Cereal Sci. Today*, **3**, 270.
67. GOLLER, H.-J. and KUBE, J. (1974). *Ind. Obst u. Gemüseverwertung*, **59**, 650.
68. DEVERO, J. E. (1973). *J. Dairy Sci.*, **56**, 286.
69. PANGBORN, R. M. and DUNKLEY, W. L. (1966). *J. Dairy Sci.*, **49**, 1.
70. ARNOTT, D. R. and BULLOCK, D. H. (1963). *Can. Dairy Ice Cream J.*, **42**(1), 27.
71. VON GENNIP, A. H. M. (1976). *Molkereitechnik*, **32**, 65.
72. WILLOCK, J. T. (1974). Brit. P. 1 369 611.
73. BURGESS, K. J. and KELLY, J. (1979). *Farm Food Rev.*, **10**(3), 78.
74. NIJPELS, H. H. (1976). *North Eur. Dairy J.*, **42**(8), 274 and **42**(10), 356.
75. HOLSINGER, V. H. (1978). *Fd Technol.*, **32**, 35.
76. MULHERIN, B., MULLEN, T., DELANEY, B. A. N. and HARPER, W. J. (1979). *New Zealand J. Dairy Sci. Technol.*, **14**, 127.
77. GUY, H. J. and EDMONDSON, L. F. (1978). *J. Dairy Sci.*, **61**, 542.
78. BLOCK, R. J. (1952). US P. 2 592 509.
79. FORSMAN, E.-S., HEIKONEN, M., KIVINIEMI, L., KREULA, M. and LINKO P. (1979). *Milchwissenschaft*, **34**(10), 618.
80. HAGGETT, T. O. R. (1976). *New Zealand J. Dairy Sci. Technol.*, **11**, 176.
81. DEMAIMAY, M., LE HENAFF, Y. and PRINTEMPS, P. (1978). *Process Biochem.*, **13**(4), 3.
82. HYRKÄS, K., VISKARI, R., LINKO, Y.-Y. and LINKO, M. (1976). *Milchwissenschaft*, **31**, 129.
83. PINOCCHIANO, T., OLSON, N. F. and RICHARDSON, T. (1980). *Adv. Biochem. Eng.*, **15**, 71.
84. HARJU, M., HEIKONEN, M., KREULA, M. and LINKO, M. (1980). In: *Food Process Engineering*, Vol. 1, Eds Linko, P., Malkki, Y., Olkku, J. and Larinkari, J., Applied Science Publishers, London, 133–6.

85. DOHAN, L. A., BARET, J. L., PAIN, S. and DELALANDE, P. (1980). In: *Food Process Engineering*, Vol. 1, Eds Linko, P., Mälkki, Y., Olkku, J. and Larinkari, J., Applied Science Publishers, London, 137–47.
86. WECKSTRÖM, L., LINKO, Y.-Y. and LINKO, P. (1980). In: *Food Process Engineering*, Vol. 1, Eds Linko, P. Mälkki, Y., Olkku, J. and Larinkari, J., Applied Science Publishers, London, 148–51.
87. MARCONI, W. (1978). In: *Biotechnology*, DECHEMA Monographien No. 1693–1703, 88–142.
88. SHAN, N. O. and NICKERSON, T. A. (1978). *J. Fd Sci.*, **43**, 1085.
89. POUTANEN, K., LINKO, Y.-Y. and LINKO, P. (1978). *Milchwissenschaft*, **33**(7), 435.
90. POUTANEN, K., LINKO, Y.-Y. and LINKO, P. (1978). *North Eur. Dairy J.*, **44**(4), 3.
91. SHAH, N. O. and NICKERSON, T. A. (1978). *J. Fd Sci.*, **43**, 1081.
92. GUY, E. J. (1978). *J. Fd Sci.* **43**, 980.
93. POMERANZ, Y., JOHNSON, J. A. and SHELLENBERGER, J. A. (1962). *J. Fd Sci.*, **27**, 350.
94. GUYRICSEK, D. M. and THOMPSON, M. P. (1976). *Cultured Dairy Prod. J.*, **13**(8), 12.
95. TAMIME, A. Y. (1978). *Cultured Dairy Prod. J.*, **13**(8), 16.
96. HARJU, M., HEIKONEN, M., KREULA, M., PAJUNEN, E. and LINKO, M. (1978). *Karjantuote*, **61**(9), 4 (in Finnish).
97. POZNASKI, S., LEMAN, J., BEDMARSKI, W., SZMELICH, W., KOWALEWSKA, J., CHADKOWSKI, M. and WIELICZKI, R. (1978). *Nahrung*, **22**, 275.
98. ROLAND, J. F. and ALM, W. L. (1975). *Biotechnol. Bioeng.*, **17**, 1443.
99. GAWEL, J. and KOSIKOWSKI, F. W. (1978). *J. Fd Sci.*, **43**, 1031.
100. POMERANZ, Y. and FINNEY, K. F. (1975). *Baker's Digest*, **49**(1), 21.
101. POMERANZ, Y., MILLER, B. S., MILLER, D. and JOHNSON, J. A. (1962). *Cereal Chem.*, **39**, 398.
102. ADACHI, S. (1958). *Nature*, **181**, 840.
103. BERNHART, F. W., GAGLIARDI, E. D., TOMARELLI, R. M. and STRIBLEY, R. C. (1965). *J. Dairy Sci.*, **48**, 399.
104. MONTGOMERY, E. M. and HUDSON, C. S. (1932). *J. Am. Chem. Soc.*, **52**, 2101.
105. PARRISH, F. W., TALLEY, F. B., ROSS, K. D., CLARK, J. and PHILLIPS, J. G. (1979). *J. Fd Sci.*, **44**, 813.
106. ROSS, K. D., PARRISH, F. W. and HUHTANEN, C. N. (1979). *39th Annual Meeting of IFT*, June 12, 1979, St. Louis, Missouri, Paper No. 280.
107. CHIRIFE, J. and FERRO FONTÁN, C. (1980). *J. Fd Sci.*, **45**, 1706.
108. GEIER, H. and KLOSTERMEYER, H. (1980). *Z. Lebensm.-Unters. u. - Forsch.*, **171**, 443.
109. GATZSCHE, L. and HAENEL, H. (1967). *Ernährungsforschung*, **16**, 641.
110. NITSCH, E. and MÜHLBÄCK, S. (1970). Ger. P. 2 002 385.
111. HAYASHIBARA K. (1970). Ger. P. 2 038 230.
112. GUTH, J. and TUMERMAN, L. (1970). US P. 3 546 206.
113. VAHERI, M. and KAUPPINEN, K. (1978). *Acta Pharm. Fennica*, **87**, 75.

114. BIRCHER, J., HAEMMERLI, U., SCOLLO-LAVIZZARI, G. and HOFFMAN, K. (1971). *Am. J. Med.*, **51**, 148.
115. MEYERHOFER, F. and PETUELY, F. (1959). *Wien klin. Wshcr.*, **71**, 865.
116. ZADOV, G. (1979). *New Zealand J. Dairy Sci. Technol*, **14**, 131.
117. RUTTLOFF, H., TÄUFEL, A., KRAUSE, W., HARNEL, A. and TÄUFEL, K. (1967). *Nahrung*, **11**, 39.
118. HOFFMAN, K., MOSSEL, D., KORUS, W. and VAN DER KRAMER, J. H. (1964). *Klin. Wschr.*, **42**, 126.
119. PETUELY, F. (1962). *Arch. Kinderheilk.*, **165**, 209.
120. MACGILLIVRAY, P. C., FINLAY, H. V. L. and BINNS, T. B. (1959). *Scot. Med. J.*, **4**, 182.
121. PETUELY, F. (1957). *Z. Kinderheilk.*, **79**, 174.
122. KARDEL, T. (1971). *Igeskr. Laeg.*, **133**(18), 889 (in Danish).
123. GRIFFIN, W. C. and LYNCH, M. J. (1972). In: *Handbook of Food Additives*, 2nd edn, Ed. Furia, T. E., CRC Press, Cleveland, Ohio, 431–55.
124. JOHNSON, J. C. (1976). *Specialized Sugars for the Food Industry* Noyes Data Corp., Park Ridge, NJ.
125. EMODI, A. (1978). *Fd Technol.*, **32**, 28.
126. HEIDEGGER, H. (1977). *Stärke*, **29**(12), 430.
127. KEARSLEY, M. W. and BIRCH, G. G. (1977). *Stärke*, **29**(12), 425.
128. HAYASHIBARA, K. (1971). Brit. P. 1 253 300.
129. HAYASHIBARA, K. (1974). Neth. P. 73 13151.
130. HAYASHIBARA, K. (1976). US P. 3 973 050.
131. SAIJONMAA, T., HEIKONEN, M., KREULA, M. and LINKO, P. (1978). *Milchwissenschaft*, **33**, 733.
132. VON VELTHUIJSEN, J. A. (1979). *J. Agr. Fd Chem.*, **27**(4), 680.
133. SAIJONMAA, T., HEIKONEN, M., KREULA, M. and LINKO, P. (1978). *Karjantuote*, **61**(10), 4 (in Finnish).
134. KARRER, P. and BÜCHI, J. (1937). *Helv. Chim. Acta*, **20**, 86.
135. BOUCHARDAT, G. (1872). *Ann. chim. phys.*, **27**(4), 68.
136. SUGIMOTO, K. (1976). US P 3 973 050.
137. MATSUO, T. (1973). *Shigaku*, **60**(6), 760.
138. TREVALON, N. V. (1963). French P. 1 317 216.
139. SCHOLNIK, F., BEN-ET, G. and SUCHARSKI, M. K. (1973). *J. Am. Oil Chem. Soc.*, **52**(7), 256.
140. VON VELTHUIJSEN, J. A., HESSEN, J. G. and KUIPERS, P. K. (1977). In: *Sucrochemistry*, Ed. Hickson, J. L., *ACS Symp. Ser.*, **41**, 136.

7

Fructose in Food Systems

L. HYVÖNEN and P. KOIVISTOINEN
*Department of Food Chemistry and Technology,
University of Helsinki, Helsinki 71, Finland*

ABSTRACT

Fructose, being a reducing monosaccharide, differs greatly from the non-reducing disaccharide, sucrose, in many of its chemical and physical properties. High solubility, hygroscopicity, sensitivity to degradation during heating, and strong browning in the Maillard-type reactions are all properties of fructose that are of importance in food systems.

Although the sweetness of crystalline fructose is nearly twice that of sucrose, the sweetness of fructose in food depends greatly on the tasting medium. Sweetening of hot beverages like coffee and tea by fructose is not advantageous, because the sweetness of fructose in these beverages is lower than that of sucrose; but in cold, slightly acid drinks the sweetness of fructose is enhanced.

In bakery products, cakes and biscuits, the high reactivity of fructose in the Maillard reaction requires that changes be made in the conventional baking procedure. Lower baking temperatures or shorter baking times are necessary. Nevertheless a well-baked fructose product is notably darker than the conventional sucrose product.

Substitution of sucrose by fructose in bakery products weight by weight does not cause significant changes in the sweetness of the product. The fructose and sucrose products are often approximately equally sweet. Differences in starch gelatinisation temperature due to sugar type (sucrose v. fructose) should be taken into account in formula adjustment since this is an important factor in determining cake structure. Fructose cakes have been reported to be smaller in volume than sucrose cakes of equal weight. In yeast-leavened doughs *fructose functions as a substrate for the yeast in a similar way to sucrose.*

The ability of fructose to enhance the flavour of fruit and berry products and, on the other hand, to cover the aftertaste of saccharin is well known. The chemical activity of fructose may cause colour faults or faster browning of the colour in, for example, strawberry jam and juice. Synergistic effects in mixtures of fructose and saccharin or cyclamate can be used to advantage especially in dietetic drinks.

The use of fructose particularly in special products as an alternative to sucrose is warranted by its partly insulin-independent metabolism, which makes it a sweetener suitable in diabetic diets. The lower cariogenicity of fructose recommends its use as a sweetener in snacks. The relative sweetening advantage of fructose is best utilised in products that contain low concentrations of sugar and in products that do not have a strong competing taste.

INTRODUCTION

Fructose has always occurred in man's food systems since it is found in the free form in almost all sweet fruits and berries. About 50% of the dry matter of honey, the oldest sweetener known to man, is fructose. Pure crystalline fructose and its use in food systems are discussed in this paper.

There are many reasons for the rising interest in sugars other than conventional sucrose: (1) Diabetics must avoid sucrose in their diet; (2) sucrose-containing snacks, especially those which remain long in the mouth, have been proven harmful to the teeth; (3) the safety of using the synthetic sweeteners, saccharin and cyclamate, has been questioned; (4) the less pleasant taste of the synthetic sweeteners (side- and aftertaste) is unpopular with consumers; (5) fructose for industrial use is more readily attainable due to the technological progress in isolating fructose and is consequently available at a lower price.

In 1980, the fructose production capacity of the main producers was estimated to be 17 000 tons of crystalline fructose and 10 500 tons of 70% liquid (w/w) fructose. Its main consumption areas are industrial use, retail sales and pharmaceutical use. The price of fructose in relation to that of sucrose varies a great deal both internationally and even nationally. In 1980 fructose was approximately twice as expensive as sucrose in Finland, while in Germany it was 3·5–4 times more expensive.

Fructose, being a reducing keto-hexose, differs considerably from the non-reducing disaccharide, sucrose, in many of its chemical and physical properties. Fructose is the most water-soluble of all the sugars. The concentration of a saturated fructose solution at 25°C is 81%, the corresponding concentration of sucrose being 67%.[1] On the other hand fructose is difficult to crystallise, because of the existence of a mixture of tautomers in solution at equilibrium.[2]

Fructose is quite hygroscopic. There is a rapid rise in its sorption isotherm at about 60% RH, at room temperature.[3] This places high demands on the packaging materials used for fructose-containing foodstuffs. On the other hand this property may help to keep bakery products, for example, moist and fresh.

The thermal treatment of hexoses in aqueous acidic solution or in molten form results in the formation of 5-hydroxymethylfurfural as the main reaction product. Fructose is, however, much more sensitive than glucose to this acid degradation.[4] The browning reactivity of fructose with amino acids is stronger than that of glucose; fructose browns more intensely in lower concentrations of amino acids and at an earlier stage of the reaction period.[5]

The high chemical reactivity of fructose probably accounts for the low stability of colour or discolouration in some fruit and berry products[6–8]

SWEETNESS

Among others, Shallenberger and Acree,[9] Lindley and Birch[10] and Shallenberger[11] have studied the saporous unit responsible for the sweet taste of fructose. It seems that only β-D-fructopyranose is sweet; the furanose forms of fructose are tasteless.

Fructose is known to crystallise only as β-D-fructopyranose. It is therefore not surprising that the crystalline fructose is the sweetest state of fructose. The sweetness of fructose solutions varies a great deal and this variation is directly related to mutarotational behaviour. The sweetness of crystalline fructose is nearly twice that of sucrose.[12,13]

In aqueous solution, the sweetness of fructose is greatly dependent on temperature. Tsuzuki and Yamazaki,[14] Fricker et al.[15]

and Hyvönen et al.[16] have reported almost identical values for the relative sweetness of fructose as a function of temperature (Table 1). Polarimetric,[17] GLC[18-20] and ^{13}C-NMR spectrometric measurements[21] showed increased amounts of the tasteless furanose forms at the mutarotational equilibrium of fructose at higher temperatures. Under corresponding conditions the sweetness of fructose is significantly decreased.

TABLE 1
RELATIVE SWEETNESS OF FRUCTOSE AT VARIOUS TEMPERATURES COMPARED WITH A 5% SUCROSE SOLUTION

Temperature (°C)	Tsuzuki and Yamazaki[14]	Fricker et al.[15]	Hyvönen et al.[16]
5	143·7	143	143
18	128·5		
22			125
25		125	
37			100
40	100	105	
50			88
60	79	81	

The relative sweetness of fructose is also dependent on concentration. Tsuzuki and Yamazaki,[14] Pangborn[22] and Schaller and Weiss[23] reported that the relative sweetness of fructose decreased with increasing concentration. In particular the sweetness of low concentrations of fructose at room and cold (5°C) temperatures is enhanced.[16, 22, 24, 25]

The tasting medium also has an effect on the relative sweetness of fructose. Slight acidity of cold fructose solutions enhances the sweetness of fructose,[26] whereas greater acidity, also of cold solutions, depresses the sweetness.[25-27]

In pear nectar, fructose is less sweet than sucrose over the concentration range 1–20%.[22] Fricker et al.[15] found that the relative sweetness of fructose is distinctly lower in cold tea (5°C) than in cold tap water, while at 60°C there is no difference in the sweetness of the fructose in the two media. In both these hot solutions the sweetness of fructose is lower than that of sucrose.

Fricker et al.[15] also noted that the greater acidity (pH 2·5) of lemon juice depressed the sweetness of fructose distinctly more than the lower acidity (pH 3·0) of grapefruit juice. Cardello et al.[25] reported a relatively large sweetening advantage of fructose over sucrose at low sugar concentrations in Kool-Aid drink (pH 2·7). This advantage disappeared, however, with increasing concentration, and in the more acidic lemon-flavoured drink (pH 2·35) no advantage of fructose over sucrose was found at any sugar concentration.

SYNERGISTIC EFFECTS IN SWEETNESS

Synergistic effects have been noted especially in mixtures of sweeteners with greatly diverging chemical structures and dissimilar relative levels of sweetness. Weickmann et al.[28] reported synergy in fructose–saccharin and fructose–cyclamate mixtures. A synergistic effect of about 20 to 30% was also found in a dextrose–fructose mixture.[29] In the sucrose–fructose experiments, only a very slight synergistic effect was observed in some of the mixtures. Yamaguchi et al.[24] defined the interrelationship between fructose and sucrose as slightly synergistic, and that between fructose and cyclamate, as well as between fructose and saccharin, as synergistic.

About 50–60% extra sweetness was perceived in aqueous fructose–saccharin solutions in the most ideal combinations of these sweeteners at the predicted isosweetness with a 5% sucrose solution.[30] In fructose–cyclamate mixtures the degree of synergism was 62–66% at the predicted isosweetness with a 5% sucrose reference, at various temperatures (8°, 25° and 50°C). At the higher sweetness level, at the predicted isosweetness with a 10% sucrose solution, the degree of synergism in fructose–cyclamate mixtures was still greater—97–98% in the optimal combination at each of the temperatures 8°, 25° and 50°C.[31, 32]

FRUCTOSE IN BEVERAGES

In slightly acidic lemonade served at 4°C fructose was considered sweeter than sucrose on an equal weight basis. The overall acceptance of the fructose-sweetened lemonade was also preferred and rated significantly higher than sucrose-sweetened lemonade.[33]

Cardello et al.,[25] however, found no difference in perceived sweetness between fructose and sucrose in lemon flavoured beverages (pH 2·35 and 2·7) at the normally used concentration of this type of drink, and Harris et al.[34] found a sucrose-sweetened orange-flavoured beverage base (pH 2·90) sweeter than a fructose-sweetened one (pH 2·95) on an equal weight basis.

Fricker et al.[15] noted that the sweetness of a fructose-sweetened grapefruit juice (pH 3·0) was enhanced as compared to a 5% sucrose-containing reference juice. However, the same workers found that in the more acidic (pH 2·5) and sweeter lemon juice (using a 15% sucrose-containing juice as the reference) the sweetness of fructose was lower than that of sucrose.

Fructose proved to be a good sweetener in a UHT-sterilised milk based chocolate drink. It was preferred at the same concentration level (4%) as sucrose in this chocolate drink. In a consumer study, 80·5% of 153 school boys and girls, aged 7–9 years, preferred the fructose-sweetened chocolate drink to a sorbitol-sweetened one.[35]

Although the sweetening of hot beverages, such as coffee and tea, with fructose is not advantageous (as the sweetness of fructose when used in this way is lower than that of sucrose) the utilisation of fructose–saccharin mixtures has proved useful as a low-calorie and dietetic sweetener for this type of drink.[36] An energy reduction of 50–70% compared to an equally sweet sucrose-sweetened coffee or tea could be achieved without a deterioration in other taste qualities.

Mixtures of fructose and saccharin and fructose and cyclamate have been found to be useful low-calorie sweeteners in soft drinks. A citrus-based soft drink can be sweetened with a mixture of fructose and saccharin to a level, which is isosweet with a 10% sucrose-sweetened beverage, without noticeable bitterness or aftertaste due to saccharin when the proportion of saccharin in the mixture is 0·4%. The energy content is 65% lower than that of a conventional sucrose-sweetened drink, while a good sweetness quality is still maintained.[37]

A cola-type soft drink requires a smaller amount of fructose–saccharin mixture than does a citrus-based soft drink to produce a sweetness equal to that of a 10% sucrose-sweetened beverage. To avoid the aftertaste of saccharin in cola-type beverages, its proportion in the mixture cannot be higher than 0·3% at the sweetness level of 10% of sucrose.[37] Both the citrus-based and cola-type test drinks containing 2·5% of fructose and 0·173% of cyclamate were judged to be as good as the corresponding sucrose-sweetened commercial soft drinks, although they had 75% less energy.[32]

FRUCTOSE IN BAKERY PRODUCTS

The quality characteristics of fat-containing cakes,[38] layer cakes,[39] white cakes,[33] vanilla cakes[25] and sugar cakes,[40] prepared with fructose have been studied.

Fructose cakes have always been smaller in volume than sucrose cakes. Bean et al.[39] noted the significant effect of the starch gelatinisation temperature on cake texture. The optimal cake texture was produced when the starch gelatinised in the range 87·5° to 92°C. The concentration of sucrose was then 56% and that of fructose 68%. This difference in the effect on gelatinisation ought to be taken into account in formula adjustment when alternative sweeteners are employed in traditional cake recipes.

The other very noticeable difference between sucrose cakes and fructose cakes is the greater browning of fructose products. If the same baking temperatures are used, the fructose cake must be removed from the oven earlier, otherwise it will burn. A properly baked fructose cake is always darker than a similar sucrose cake. It should also be noted that fructose and sucrose cakes usually taste equally sweet despite the greater sweetness of fructose in aqueous solutions at room temperature.[25, 33, 40]

In biscuits sucrose was judged to be significantly sweeter than fructose.[33, 40] A greater degree of browning and less spreading was noted in the fructose biscuits. The textures of the sucrose biscuits and the fructose biscuits were different; fructose biscuits were characterised as chewy while the sucrose biscuits were crisp in texture. Hardy et al.[33] noted that fructose biscuits stored in a plastic bag at room temperature stayed moist and chewy for several weeks. This has been explained by the high absorptive power of fructose for water in a saturated atmosphere.[41]

In yeast-leavened doughs fructose functions as a substrate for the yeast in a similar way to sucrose.[42] In a sensory evaluation fructose buns were judged to be slightly darker than sucrose buns. Fructose buns were slightly smaller in volume than sucrose buns. Generally speaking, however, sucrose and fructose buns are very similar.[43]

FRUCTOSE IN JAMS, JELLIES AND PUDDINGS

In the preparation of strawberry jam, fructose behaved very similarly to sucrose although during storage the colour of fructose jam deteriorated sooner than that of sucrose jam.[8] Ellala[6] reported a purple colouration, even in freshly prepared strawberry jam, when a solution of 70% of non-crystallised fructose was used, but not when crystalline fructose was used.

Kawabata et al.[44] have reported data, which show that fructose and sucrose behave differently when used in HM-pectin jellies. In LM-pectin jellies the effects of fructose and sucrose on the texture were very similar. At equal sweetness the sensory preference for the fructose and sucrose jellies was not significantly different.

According to Hardy et al.[33] vanilla puddings sweetened with equal weights of sucrose and fructose did not differ significantly in sweetness, flavour, texture or overall acceptance.

FRUCTOSE IN ICE CREAM

Sugar has an important effect on the texture and melting of ice cream, both of which are very important quality characteristics. The added sugar depresses the freezing point of ice cream, i.e. the higher the sugar content, the lower the melting point of ice cream. When the depression is a function of molecular weight,[45] fructose with a lower molecular weight (mol. wt = 180) depresses the freezing point of ice cream more than an equal amount of sucrose (mol. wt = 342).

Steinsholt[46] reported, however, that only dextrose had a considerable effect on the freezing point of ice cream in an experiment where 0, 10, 20 and 30% of the sucrose was replaced by dextrose, fructose and glucose, respectively. The use of dextrose caused a softer ice cream at $-21°C$. He recommended replacing only 25% of the sucrose by fructose in the ice cream in order to avoid any negative effects upon the quality of the ice cream.

In our study in which sucrose was replaced by fructose on a weight basis in ice cream there was no significant difference in the appearances, textures or tastes of the sucrose- and fructose-sweetened ice creams. After three months' storage at $-25°C$, the melting of fructose ice cream was still judged to be satisfactory while the melting of sucrose ice cream stored in the same freezer was judged to be unsatisfactory.[47]

FRUCTOSE IN CHOCOLATE

The use of fructose in the manufacture of chocolate demands special treatment in different chocolate processes. Fructose crystals pulverise less easily under shock or shear than do sucrose crystals. The crystals, especially the ground crystals, have a natural tendency to agglomerate by adhesion, which is accentuated by their hygroscopicity. The working temperature during conching is important in determining whether agglomeration of the particles of fructose will occur, a temperature of about 40°C being desirable. The chemical reactions of fructose with the other constituents take place easily in the conching process.[48]

Manufacturing should be carried out as rapidly as possible to reduce the chance of moisture pick-up and to avoid agglomeration. Moulding or enrobing should be done immediately after conching and tempering, thus avoiding storage of the chocolate.[48]

Liquid centres for casting and injection are easily made from fructose. Fructose is suitable for this application in that; (a) it is very soluble, (b) it forms high concentrations without recrystallisation, (c) it does not have an excessively high viscosity, and (d) it lacks a tendency to penetrate the chocolate shell.[48]

FRUCTOSE AS YOGURT SWEETENER

Fructose was found to be sweeter than sucrose in yogurt.[49,50] Salminen and Branen[49] reported that yogurt sweetened with fructose was evaluated as slightly more tart, but fresher than sucrose-sweetened yogurt. Hyvönen and Slotte[50] noted the good flavour of the fructose-sweetened yogurt, which was judged to be better after two weeks' storage than when fresh. The appearance and texture of fructose yogurt were found to be similar to those of sucrose-sweetened yogurt.

BENEFITS OF FRUCTOSE IN DIETETIC FOODS

The use of fructose particularly in special products as an alternative to sucrose is warranted by its partly insulin-independent metabolism,[51] which makes it a sweetener suitable in diabetic diets. The lower cariogenicity of fructose[52] recommends its use as a sweetener in snacks. However, the use of fructose in foods which are heat-treated may cause some problems and may require changes in conventional procedures.

The relative sweetening advantage of fructose is best utilised in products containing low concentrations of sugar, such as slightly acidic drinks consumed cold, and in products not having a strong competing taste. The sweetness advantage of fructose over sucrose is often lost in solid food systems, however. Therefore, a surer and more pleasant means of energy reduction in dietetic foods is achieved by decreasing the size or the number of servings, rather than by reducing the amount of fructose in solid food systems.

REFERENCES

1. GRAEFE, G. (1975). *Stärke*, **27**(5), 160–9.
2. SHALLENBERGER, R. S. and BIRCH, G. G. (1975). *Sugar Chemistry*, Avi Publishing Company, Inc. Westport, Connecticut, 46–88.
3. VON HERTZEN, G. and LINDQVIST, C. (1980). In: *Carbohydrate Sweeteners in Foods and Nutrition*. Eds Koivistoinen, P. and Hyvönen, L. Academic Press, London, 127–49.
4. THEANDER, O. (1980). In: *Carbohydrate Sweeteners in Foods and Nutrition*. Eds Koivistoinen, P. and Hyvönen, L. Academic Press, London, 185–99.
5. KATO, H., YAMAMOTO, M. and FUJIMAKI, M. (1969). *Agr. Biol. Chem.*, **33**(6), 939–48.
6. ELLALA, A. (1971). 'Factors Influencing the Colour of the Fructose-Sweetened Strawberry jam,' Research report, University of Helsinki (in Finnish).
7. EL-KADY, S. and AMMAR, K. (1977). *J. Agr. Res.*, Tanta University, **3**(1), 99–111.
8. HYVÖNEN, L. and TÖRMÄ, R. (1981). *J. Fd. Sci.*, (in press).
9. SHALLENBERGER, R. S. and ACREE, T. E. (1969). *J. Agr. Food Chem.*, **17**(4), 701–3.
10. LINDLEY, M. G. and BIRCH, G. C. (1975). *J. Sci. Fd. Agric.*, **26**, 117–24.
11. SHALLENBERGER, R. S. (1978). *Pure & Appl. Chem.*, **50**, 1409–20.
12. SHALLENBERGER, R. S. (1963). *J. Fd. Sci.*, **28**, 584–9.
13. HYVÖNEN, L. and RATILAINEN, A. (1974). 'Sweetness of the Crystalline

and Freshly Dissolved Fructose,' Research report, EKT-series 351, University of Helsinki (in Finnish).
14. TSUZUKI, Y. and YAMAZAKI, J. (1953). *J. Biochem.* **323**, 525–31.
15. FRICKER, A., PROCHAZKA, E. and GUTSCHMIDT, J. (1973). *Lebensm. -Wiss. u. -Technol.*, **6**, 63–5.
16. HYVÖNEN, L., KURKELA, R., KOIVISTOINEN, P. and MERIMAA, P. (1977). *Lebensm. -Wiss. u. -Technol.*, **10**, 316–20.
17. HYVÖNEN, L., VARO, P. and KOIVISTOINEN, P. (1977). *J. Fd. Sci.*, **42**, 652–3.
18. SHALLENBERGER, R. S. (1973). *Adv. in Chem. Series*, **117**, 256–63.
19. MAUCH, W. and FARHOUDI, E. O. (1976). *Z. Zuckerind.*, **26**, 766–71.
20. HYVÖNEN, L., VARO, P. and KOIVISTOINEN, P. (1977). *J. Fd. Sci.*, **42**, 654–6.
21. HYVÖNEN, L., VARO, P. and KOIVISTOINEN, P. (1977). *J. Fd. Sci.*, **42**, 657–59.
22. PANGBORN, R. M. (1963). *J. Fd. Sci.*, **28**, 726–33.
23. SCHALLER, A. and WEISS, J. (1979). *Confructa*, **24**, 85–93.
24. YAMAGUCHI, S., YOSHIKAWA, T., IKEDA, S. and NINOMIYA, I. (1970). *Agr. Biol. Chem.*, **34**, 181–6.
25. CARDELLO, A. V., HUNT, D. and MANN, B. (1979). *J. Fd. Sci.*, **44**, 748–51.
26. HYVÖNEN, L., KURKELA, R., KOIVISTOINEN, P. and ALA-KULJU, M.-L. (1978). *Lebensm. -Wiss. u. -Technol.*, **11**, 11–14.
27. PANGBORN, R. M. (1965). In: *Proceedings of the 1st International Congress of Food Science and Technology*, Vol. III, (London) Ed. Leitch, C. M., Gordon and Breech Science Publishers, New York, 291–305.
28. WEICKMANN, F., WEICKMANN, H., FINCKE, K., WEICKMANN, F. A. and HUBER, B. (1969). Ger. P. 1961769. Deutsche Laevosan-Gesellschaft G. F. Boehringer und Söhne GmbH und Co. KG., Mannheim.
29. STONE, H. and OLIVER, S. M. (1969). *J Fd. Sci.*, **34**, 215–22.
30. HYVÖNEN, L., KURKELA, R., KOIVISTOINEN, P. and RATILAINEN, A. (1978). *J. Fd. Sci.*, **43**, 251–4.
31. SIPILÄ, L. (1977). 'Replacement of Sucrose in Foods,' Research report, EKT-series 414. University of Helsinki (in Finnish).
32. HYVÖNEN, L. and SIPILÄ, L. (1977). 'Effects of Temperature and Mixture Combinations on the Synergism between Fructose and Cyclamate, and Xylitol and Cyclamate, and the Application of the Mixtures of Sweeteners in Foods,' Research report, EKT-series 421, University of Helsinki (in Finnish).
33. HARDY, S. L., BRENNAND, C. P. and WYSE, B. W. (1979). *J. Am. Diet. Assoc.*, **74**, 41–6.
34. HARRIS, N. E., RUBINO, J. A. and MCNUTT, J. W. (1978). *Food Process. Ind.*, (February) 28–9.
35. HYVÖNEN, L. and ESPO, A. Unpublished information.
36. HYVÖNEN, L., KURKELA, R., KOIVISTOINEN, P. and RATILAINEN, A. (1978). *J. Fd. Sci.*, **43**, 1577–79, 1584.

37. HYVÖNEN, L., KOIVISTOINEN, P. and RATILAINEN, A. (1978). *J. Fd. Sci.*, **43**, 1580–4.
38. THOMPSON, C. M., FUNK, K., SCHEMMEL, R. and MICKELSEN, O. (1974). *Ecol. Food. Nutr.*, **3**, 231–6.
39. BEAN, N. M., YAMAZAKI, W. T. and DONELSON, D. H. (1978). *Cereal Chem.*, **55**(6), 945–52.
40. HYVÖNEN, L. and ESPO, A. (1981). *Lebensm.-Wiss. u.-Technol.*, (in press).
41. AUDU, T. O. K., LONCIN, M. and WEISSER, H. (1978). *Lebensm. -Wiss. u. -Technol.*, **11**, 31–4.
42. VARO, P., WESTERMARCK-ROSENDAHL, C., HYVÖNEN, L. and KOIVISTOINEN, P. (1979). *Lebensm. -Wiss. u. -Technol.*, **12**, 153–6.
43. HYVÖNEN, L. and ESPO, A. (1981). *Lebensm. -Wiss. u. -Technol.*, (in press).
44. KAWABATA, A., SAWAYAMA, S. and KOTOBUKI, S. (1976). *Japanese J. Nutr. (Eiyogaku Zasshi)*, **31**, 3–10.
45. NICOL, W. M. (1980). In: *Carbohydrate Sweeteners in Foods and Nutrition*, Eds Koivistoinen, P. and Hyvönen, L. Academic Press, London, 151–62.
46. STEINSHOLT, K. (1974). *Meierposten*, **63**(7), 136–141; **63**(8), 169–76; **63**(9), 195–8; **63**(10), 228–31.
47. HYVÖNEN, L. and TÖRMÄ, R. Unpublished information.
48. ZIMMERMAN, M. (1974). *Manuf. Confec.*, (August) 39–48.
49. SALMINEN, S. J. and BRANEN, A. L. (1978). 'Xylitol and Fructose as Yoghurt Sweeteners,' Paper presented at the 38th IFT Meeting, June 4–7, Dallas.
50. HYVÖNEN, L. and SLOTTE, M. (1981). *J. Fd. Sci.*, (in press).
51. YLIKAHRI, R. H. and PELKONEN, R. (1980). In: *Carbohydrate Sweeteners in Foods and Nutrition*, Eds Koivistoinen, P. and Hyvönen, L. Academic Press, London, 15–35.
52. SCHEININ, A., MÄKINEN, K. and YLITALO, K. (1975). *Acta Odont. Scand.*, **33**, 67–86.

# 8

Hydrogenated Glucose Syrups, Sorbitol, Mannitol and Xylitol

P. J. SICARD
Roquette Frères,
62136 Lestrem,
France

ABSTRACT

In this paper dealing with derivatives obtained by the catalytic hydrogenation of mono- or oligosaccharides, only those products in regular use in the food industries will be taken into consideration. For this reason Palatinit,® lactitol and glucosylsorbitol will be omitted as they can be considered as being still in the development stage. Despite the natural occurrence of hydrogenated sugars in plants and algae, where they are present as metabolites, their concentration is rarely high enough to allow direct extraction for industrial purposes.

Thus, di- or polysaccharides are generally used as industrial sources of these products; their chemical or enzymatic conversion leads to raw material for hydrogenation to one of the derivatives mentioned in this paper.

With natural sugars, the loss of the reducing function can confer some of the following properties:

1. increased chemical stability,
2. resistance to enzymatic or microbiological action,
3. special behaviour towards digestive hormonal systems,
4. pharmacodynamical activity,
5. increased affinity for water,
6. reduced tendency to crystallisation,

®Registered Trade Mark of Süddeutsche Zuckes—AG, Grünstadt, Germany.

7. improvement to technological qualities,
8. increase of sweetening power.

The special characteristics of sorbitol, mannitol, xylitol and hydrogenated glucose syrup enable them to find different applications in traditional food products, where they can be substituted for sucrose and glucose syrups. Among the hydrogenated glucose syrups, Lycasin® presents a special case insofar as it combines a good sweetening power with textural qualities. Lycasin 80/55 which presents the most developed form of this kind of product is moreover characterised by its non-cariogenicity and good digestive tolerance.

INTRODUCTION

The natural occurrence of hexitols and pentitols has been known for more than a century. As early as 1872, the French chemist Boussingault was able to characterise sorbitol as a component of the berries of the mountain ash tree (*Sorbus aucuparia*),[1] from which it derives its name. Xylitol was prepared in 1891, by Fischer[2] by the catalytic reduction of D-xylose, while its natural occurrence was simultaneously demonstrated in various fruits and vegetables.

However, with the exception of D-mannitol which can be extracted from various seaweeds,[3] the concentration of sugar alcohols in plants is generally too low to permit commercial extraction. Thus chemical synthesis remains the only route, through hydrogenation of the corresponding reducing sugar, to these products on an industrial scale.

In this paper, it was decided to consider only those hydrogenated sugars which are of established industrial importance; accordingly products such as lactitol, glucosylsorbitol and Palatinit are omitted from this review. The physico-chemical differences which exist between the sugar alcohols and the sugars, will first be considered.

In food applications, the main reason for hydrogenating reducing sugars is to increase chemical stability and affinity for water

® Registered Trade Mark of Roquette Frères, Lestrem, France.

without altering sweetening power. Moreover, this chemical modification of sugars generally lowers their tendency to crystallise and also makes their metabolism non-insulin-dependent and strongly depresses their cariogenicity.

The first hydrogenated sugar to have been manufactured was sorbitol, in the 1930s. Originally this compound was used for vitamin C synthesis, and subsequently for the manufacture of non-ionic surfactants. It is used as a humectant and anti-caking agent in baking in the USA, as an additive to cosmetics and as a polyol in the synthesis of rigid polyurethane foams.

On the chemical market sorbitol was followed by mannitol and then hydrogenated glucose syrups including Lycasin, and finally, xylitol. To give a precise idea of the importance of these various sugars relative to the traditional sweeteners, sucrose and glucose, their respective annual production and average price are listed in Table 1.

TABLE 1
COMPARISON OF THE WORLD ANNUAL PRODUCTION AND CURRENT PRICE OF VARIOUS NATURAL AND REDUCED SUGARS

	World production (1980; metric tonnes)	Current selling prices (relative to sucrose)
Sucrose	$0.9-1.0 \times 10^8$	1.0
Glucose and isoglucose	10^7	0.9
Liquid sorbitol (70% w/w)[a]	360 000	1.4
Crystalline sorbitol[a]	55 000	2.3
Crystalline mannitol	10 000	4.0
Lycasin syrup (75% dry wt)	3 000	2.0
Crystalline xylitol	2 000	10

[a] The actual production of sorbitol is higher than quoted here, owing to its captive use for vitamin C synthesis.

As can be seen from Table 1 the overall production of hydrogenated sugars is less than 1% of sucrose production, clearly indicating that there is no possible competition between these sweeteners.

In the manufacture of reduced sugars all processes have in common a hydrogenation step carried out under pressure and in the presence of Raney nickel (Fig. 1).

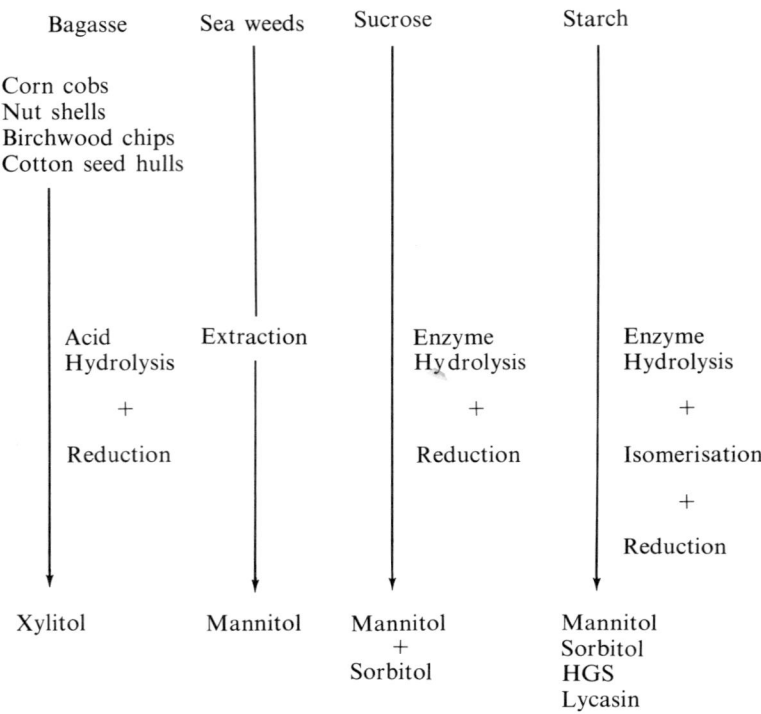

FIG. 1. Industrial origin of sugar alcohols and hydrogenated glucose syrups (HGS).

PRODUCTION OF XYLITOL: $C_5H_{12}O_5$

$$\begin{array}{c} CH_2OH \\ | \\ HCOH \\ | \\ HOCH \\ | \\ HCOH \\ | \\ CH_2OH \end{array}$$

As already mentioned, xylitol occurs naturally in various plants, but at too low a concentration to allow industrial extraction; yellow plums or greengages (*Prunus domestica italia*) are the richest source with a maximum content of 0·9% (w/w). This explains why much interest has been focussed on its precursor, D-xylose, which is widely distributed as a polymer, xylan, a major component of hemicelluloses.

The most attractive sources of D-xylose are:[4]

1. A gum obtained by the alkaline extraction of the pericarp of maize kernels, whose D-xylose content reaches 48–54%.
2. Corn cobs, containing 20–25% of D-xylose.
3. Seed husks of *Plantago arenaria* (62% D-xylose), *Plantago lanceolata* (72% D-xylose), *Plantago ovata* (46% D-xylose) and *Linum usitatissimum* (25% D-xylose).
4. Dulsan, a polymer extracted from the seaweed *Rhodymenia palmata*.
5. Nut shells.
6. Cotton seed hulls.
7. Birch wood chips.
8. Bagasse.

The hemicellulose fraction containing D-xylose can be separated by extraction with alkali or directly hydrolysed in the presence of cellulose. Hydrolysis is carried out under acid conditions using sulphuric, phosphoric or oxalic acid. According to the process used, D-xylose can be purified by crystallisation prior to its hydrogenation or hydrogenated without purification, xylitol being subsequently purified by crystallisation.

PRODUCTION OF D-MANNITOL: $C_6H_{14}O_6$

$$\begin{array}{c} CH_2OH \\ | \\ HOCH \\ | \\ HOCH \\ | \\ HCOH \\ | \\ HCOH \\ | \\ CH_2OH \end{array}$$

D-Mannitol, the name of which is derived from its occurence in manna, a sweet exudate from the ash tree (*Fraxinus ornus*), is relatively abundant in seaweeds such as *Laminaria digitata*, which have been used as raw materials for its production. According to the season of harvesting, the dry substance present in the upper part of the seaweeds contains up to 10% of D-mannitol. Thus, it is possible to couple the production of alginic acid with the extraction of D-mannitol as a by-product.

This process is currently used in China, though it appears to be limited, due to the large amount of seaweed that has to be processed to obtain significant quantities of D-mannitol.

Invert sugar is normally used as a raw material for the production of D-mannitol, reduction leading to a mixture of sorbitol (75%) and mannitol (25%). Mannitol readily crystallises from solution owing to its low solubility in water. The mother liquor, which contains predominantly sorbitol with between 8–10% D-mannitol, is concentrated to 70% solids, and used in various applications.

Recently, starch has been used as a raw material for D-mannitol production (Fig. 2). The first step of the process consists in obtaining a 97 DE hydrolysate. This is followed by one of two possible processes:

1. Dextrose is purified by crystallisation, and is then epimerised before hydrogenation to 25–30% of D-mannose in the presence of ammonium molybdate.
2. The hydrolysate is enzymically isomerised using glucose isomerase, to give a mixture of 44% fructose + 50% glucose + 6% polysaccharides, roughly equivalent to invert sugar, and catalytically hydrogenated.

PRODUCTION OF SORBITOL: $C_6H_{14}O_6$

$$\begin{array}{c} CH_2OH \\ | \\ HCOH \\ | \\ HOCH \\ | \\ HCOH \\ | \\ HCOH \\ | \\ CH_2OH \end{array}$$

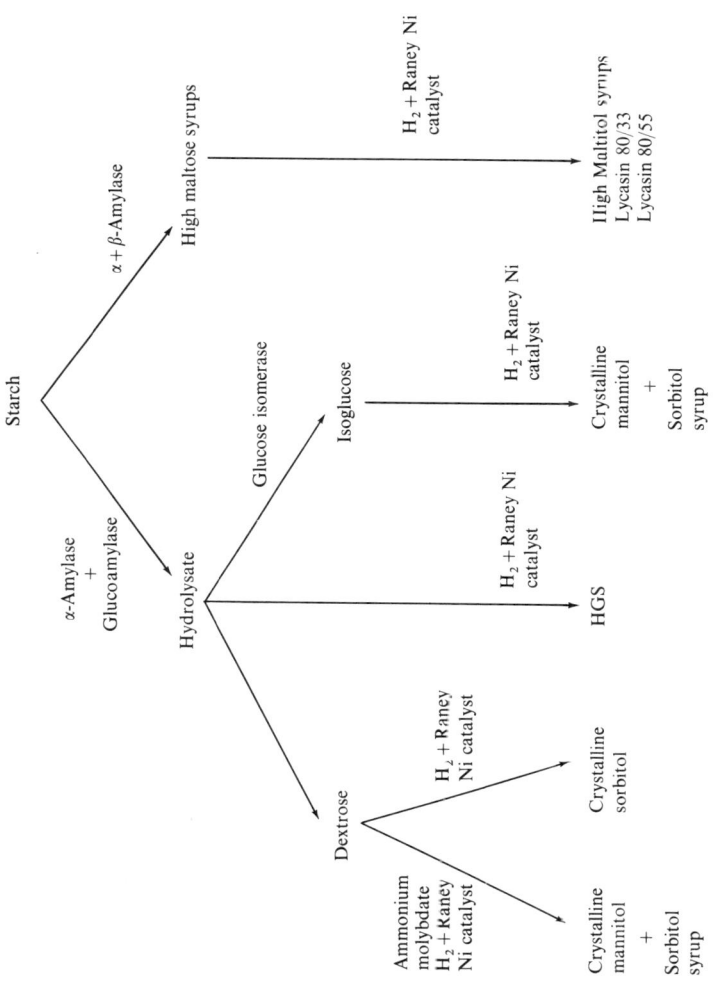

FIG. 2. Hydrogenated sugars from starch.

The method of manufacture of sorbitol depends on the state of purity required since the sorbitol content of commercial preparations can vary widely.

The manufacture of sorbitol syrups can make use of several raw materials: glucose syrups, invert sugar, isoglucose or even epimerised dextrose where mannitol is also produced (Fig. 2).

The simultaneous formation of mannitol and sorbitol in the hydrogenation of invert sugar provides an explanation for the specification of the *Food Chemicals Codex*[5] for sorbitol which allows a content of up to 9% of non-sorbitol—the usual percentage of mannitol after its separation by crystallisation. Pure sorbitol is obtained from pure dextrose by hydrogenation when purity of the final product exceeds 99·8%.

Owing to its hygroscopicity, sorbitol cannot be crystallised directly from an aqueous medium except by special techniques. Since 1972, the French firm Roquette[6] has used a crystallisation process in which a crystalline product with a stable gamma structure is obtained. This structure is characterised by a melting point higher than 96°C (the unstable beta-structure melting around 92°C).[7]

Characterisation of the crystalline purity of sorbitol may be made by X-ray spectrometry or differential thermal analysis.

PRODUCTION OF HYDROGENATED GLUCOSE SYRUPS AND LYCASIN

Apart from simple polyols resulting from the reduction of well-defined sugars, much attention has been devoted since the early 1960s to more complex products arising from the catalytic hydrogenation of hydrolysed starch syrups of various oligosaccharide composition.

The Swedish firm Lyckeby originally studied the new family of compounds obtained by the hydrogenation of potato starch hydrolysates and patented both their preparation and applications.[8] Subsequently the patents on the so-called 'Lycasin' syrups have been acquired by Roquette, enabling them to develop new types of Lycasin by substituting corn starch for potato starch and enzyme hydrolysis for acid hydrolysis. Currently, two types of Lycasin are manufactured by Roquette, Lycasin 80/33 and Lycasin 80/55. The specific properties of these products will be examined in a later section.

Undoubtedly, Lycasin 80/55 represents the most sophisticated type of hydrogenated glucose syrup. Owing to its high maltitol content, it has sufficient sweetening power not to require mixing with sweeter products; moreover, it is non-cariogenic and shows texturing properties which enable it to be used alone in the manufacture of hard boiled candies.

Normally, Lycasin 80/55 is obtained as a syrup with a dry matter content of 75%; however, it is possible to obtain it as an anhydrous powder by evaporating the syrup to a glass, which is powdered after cooling with solid carbon dioxide.

PHYSICO-CHEMICAL PROPERTIES OF HYDROGENATED SUGARS

Some of the more important physico-chemical properties of xylitol, sorbitol and mannitol are listed in Table 2. No equivalent figures are available for maltitol.

TABLE 2
PHYSICO-CHEMICAL PROPERTIES OF SIMPLE HYDROGENATED SUGARS

	Xylitol	Sorbitol	Mannitol
Chemical formula	$C_5H_{12}O_5$	$C_6H_{14}O_6$	$C_6H_{14}O_6$
Mol. wt.	152·15	182·17	182·17
Melting point (°C)	93–94·5	96–97	165–168
Heat of solution at 25°C (cal/g)	−36·61	−28	−28·9
Solubility in water at 25°C (g/100 g water)	200	235	22
Calorific content (k cal/g)	4	4	4

The relatively high heat of solution (measured at 25°C) of the three polyols accounts for their 'cooling effect' much appreciated in confectionery. The effect with mannitol is less pronounced than with sorbitol and xylitol, even though its heat of solution is identical to that of sorbitol, owing to its lower water solubility.

Hydrogenation of a reducing sugar does not necessarily lead to increased water solubility and a reduction in melting point as is often quoted, though the two properties vary in the same direction (Table 3).

TABLE 3
COMPARISON OF THE WATER SOLUBILITY AND MELTING POINT OF VARIOUS NATURAL SUGARS AND THEIR CORRESPONDING ALCOHOLS

	Melting point (°C)	Water solubility at 25°C (g/100g H_2O)
Galactose	167	200
Galactitol	188–189	3·3
Glucose	146	100
Sorbitol	98	235
Fructose	103–105	400
Mannitol	166–168	22
Mannose	133	250
Xylose	90–91	125
Xylitol	94	200

The caloric value of the sugar alcohols is unchanged on reduction of sugars, since they remain metabolisable. However in the literature, values below 4 are sometimes quoted for mannitol. This is a consequence of the low metabolic capacity of human subjects for mannitol, which when exceeded causes mannitol to be excreted unchanged in the urine, giving an apparent reduced caloric equivalent.

SWEETNESS OF HYDROGENATED SUGARS

The sweetness of sugars and polyols related to sucrose (Table 4) shows that generally there is not a great difference in sweetness between sugars and the corresponding sugar alcohols. Obviously the quoted figures must be considered with some caution, due to their variability resulting from the way in which they are determined.

COMPARISON OF THE PROPERTIES OF LYCASIN-TYPE PRODUCTS

The oligosaccharide composition of the three most commonly produced Lycasin syrups is given in Table 5.

TABLE 4
RELATIVE SWEETNESS OF VARIOUS SUGARS AND OF THE CORRESPONDING SUGAR ALCOHOLS

Sugar	Sugar alcohol	Sweetness relative to sucrose
Sucrose		100
Glucose		70
	Sorbitol	60
Fructose		150
Mannose		60
	Mannitol	50
Xylose		70
	Xylitol	90
Maltose		40
	Maltitol	90
Starch hydrolysate		25–60
	Lycasin 80/33	40
	Lycasin 80/55	55

TABLE 5
COMPARISON OF THE OLIGOSACCHARIDE COMPOSITION (%) OF VARIOUS LYCASIN SYRUPS

	'Swedish' Lycasin	Lycasin 80/33	Lycasin 80/55
DP 1	8	5	6
DP 2	8	25	52
DP 3		18	18
DP 4	22	15	1·5
DP 5		5·5	2
DP 6		3·5	2·7
DP 7		2·5	3·4
DP 8		1·0	2·2
DP 9	62	1·0	1·0
DP 10		0·5	1·0
20 > DP > 10		2·0	9·2
DP > 20		21·0	0·8

The original 'Swedish' Lycasin, obtained by an acid hydrolysis of potato starch followed by catalytic reduction, and the first to be marketed, is characterised by a low sweetening power and a high content of oligosaccharides.

When Roquette became involved with Lycasin, the substitution of corn starch for potato starch resulted in the product designated as Lycasin 80/33 and characterised by texturing properties and limited sweetness.

Further development was directed towards achieving the following properties:

1. non-cariogenicity,
2. improved sweetness,
3. textural properties,
4. low free sorbitol content

The utilisation of manufacturing processes using ingredients accepted as safe within the food industry with due consideration for the legislative situation internationally, was also desirable.

In order to increase the sweetness of Lycasin, a high content of maltitol is required. This can be obtained by submitting the corn starch hydrolysate to the action of the enzyme β-amylase to produce a high maltose syrup.

To give the final product its non-cariogenicity the various polysaccharide compositions, which resulted from a combination of several enzymes and a modification of reaction parameters, were submitted to dental evaluation using the pH-telemetry system devised by Imfeld and Mühlemann.[9] It was found that non-cariogenicity could only be obtained if the final product contained less than 3% of oligosaccharides with a degree of polymerisation higher than 20.

The hydrogenated corn starch hydrolysates which comprise Lycasin 80/55 represent the optimal composition. The process and applications of the product are covered by patents.[10] Lycasin 80/55 is characterised by a high stability to heat (withstanding temperatures up to 180°C) and a high stability towards microbial degradation.

METABOLISM OF HYDROGENATED SUGARS

Reduction of natural sugars will modify their metabolic behaviour in addition to their physico-chemical properties. Any carbohydrate which is not normally present in the human diet (other than, for example, glucose, fructose, sucrose, lactose, starch), the metabolism of which follows classical enzymatic routes, will only be metabolised if it is converted in the alimentary tract into compounds which can enter normal metabolic pathways, where it will undergo oxidative reactions leading to the production of energy.

Alternatively the original substance may be absorbed through the intestinal wall before entering the metabolic pathway. Since only D-glucose, D-galactose and D-allose are known to be actively transported,[11] in order to cross the intestinal membrane, other substances such as sorbitol must be absorbed by passive diffusion.

Generally, the metabolism of hydrogenated sugar involves well-known metabolic pathways such as the Embden–Meyerhof–Parnas glycolytic pathway, the Horecker–Racker pentose phosphate pathway, the glucuronate–xylulose Touster's cycle and the Krebs' cycle, to achieve the degradation of pyruvate when it appears as an intermediate in sugars degradation.

METABOLISM OF SORBITOL

Sorbitol enters the metabolism of higher organisms by the fructose pathway (Fig. 3). Although it is not one of the main human metabolites, sorbitol is found in several tissues which can catalyse the reduction of D-glucose to sorbitol through the action of aldose reductase (EC 1.1.1.21):[12]

$$\text{D-glucose} + \text{NADPH} \longrightarrow \text{sorbitol} + \text{NADP}^+$$

This explains why sorbitol can be found in certain tissues or fluids, even in the absence of any exogenous administration.

The first metabolic transformation undergone by sorbitol is its oxidation to D-fructose, catalysed by the enzyme sorbitol dehydrogenase, (L-iditol: NAD^+ 5-oxydoreductase, EC 1.1.1.14):

$$\text{sorbitol} + \text{NAD}^+ \longrightarrow \text{D-fructose} + \text{NADH}$$

This enzyme occurs mainly in the liver, prostate gland and kidneys, with relative activities of 5·73/1·37/1·24.

Sorbitol oxidation and the subsequent metabolism of D-fructose are hepatic reactions which enable these compounds to join normal glycolytic pathways through non-insulin-dependent mechanisms. In fact, the activity of the enzymes involved in these transformations (L-iditol dehydrogenase (EC 1.1.1.14), fructokinase (EC 2.7.1.4) and hepatic aldolase (EC 4.1.2.13)) is not controlled by insulin. On the other hand, the level of glucokinase (EC 2.7.1.2)

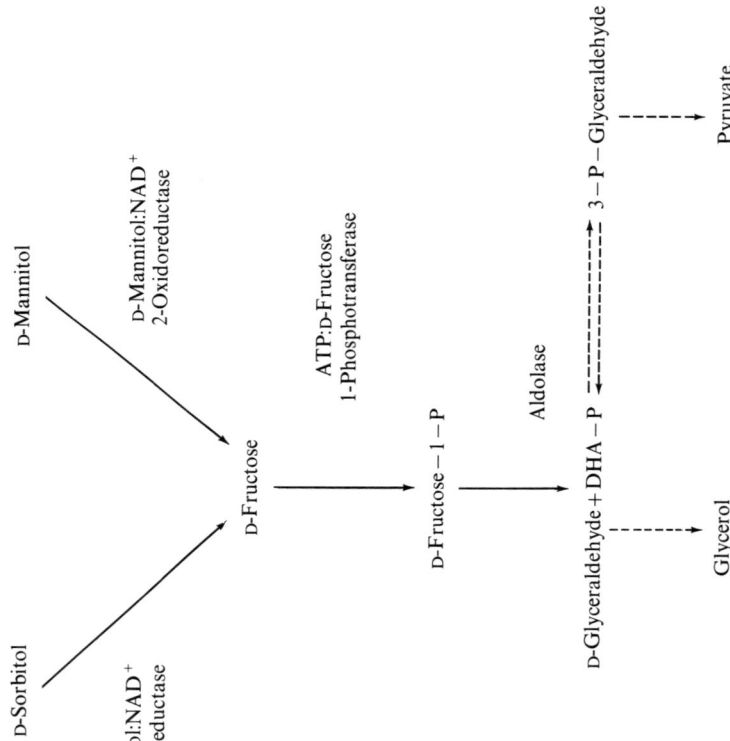

Fig. 3. Metabolism of sorbitol and mannitol.

which is insulin-dependent may be inadequate in diabetics with insulin deficiency.

The transformation of sorbitol to D-fructose gives it access to several metabolic pathways. Two examples are given below:

(i) D-Fructose can enter the Embden–Meyerhof–Parnas glycolytic pathway, through its phosphorylation to fructose-6-phosphate, catalysed by hexokinase (EC 2.7.1.1). This transformation also gives it access to the pentose phosphate cycle.

In order to establish a balance of energy production during sorbitol metabolism, it is necessary to consider all transformations which enable it to enter the Krebs' cycle (Fig. 4). If only the reactions yielding or consuming ATP and oxido-reduction coenzymes are considered, the metabolic distance between sorbitol and pyruvate, is equivalent to 2 ATP + 3 NADH (Fig. 5).

The oxidation of pyruvate in the Krebs' cycle is described by the following stoichiometry:

$$CH_3COCOOH + 5/2 O_2 + 15 ADP + P_i \longrightarrow 3CO_2 + 17H_2O + 15ATP$$

Assuming that 1 mol NADH is equivalent to 3 mol ATP, the respective theoretical values of 38 mol ATP and 41 mol ATP produced by the overall oxidation of 1 mol of D-glucose and 1 mol of sorbitol are obtained:

$$C_6H_{12}O_6 \longrightarrow 6CO_2 + 6H_2O + 32ATP + 2NADH$$
$$C_6H_{14}O_6 \longrightarrow 6CO_2 + 7H_2O + 32ATP + 3NADH$$

In practice, for D-glucose, and consequently for sorbitol, the number of moles of ATP generated by complete oxidation is slightly less. This results from exchange phenomena occurring at the level of the mitochondrial membrane, during which some energy is lost, so that, for D-glucose, only 35·5 mol ATP are recovered instead of 38.[13]

If instead of normal glycolysis, fructose-6-phosphate enters the pentose phosphate cycle, the overall oxidation reaction becomes:

$$C_6H_{12}O_6 + 7H_2O + 12NADP^+ + ATP \longrightarrow 6CO_2 + 12NADPH + 12H^+ + ADP + P_i$$

As 1 mol NADH is generated by the oxidation of sorbitol to D-

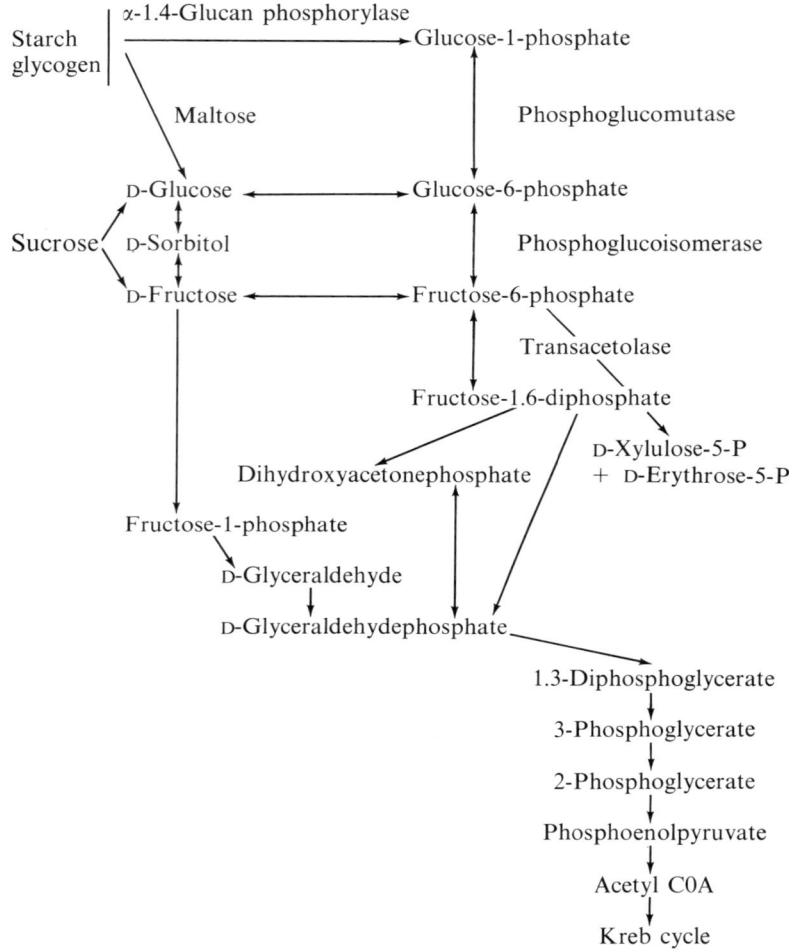

FIG. 4. Transformation of carbohydrates, sorbitol and D-fructose to acetyl-COA.

fructose, potentially 38 mol ATP are produced during this type of degradation.

(ii) D-Fructose is transformed to fructose-1-phosphate in the presence of hepatic fructokinase (EC 2.7.1.4), which acts only when enough Mg^{2+} is present; then fructose-1-phosphate is degraded under the influence of a second hepatic enzyme, 1-phosphofructaldolase (EC 4.1.2.13) (Fig. 6).

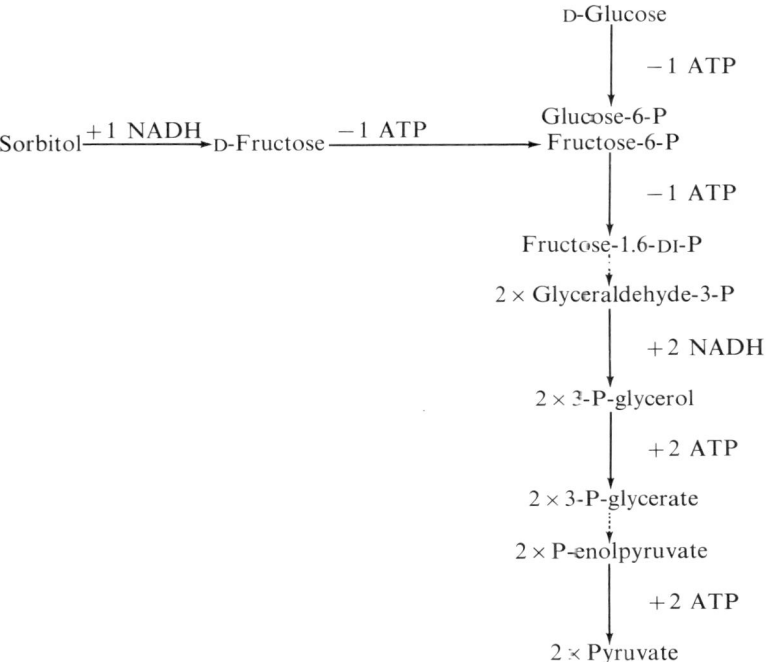

FIG. 5. Energetic balance of the transformation of sorbitol to pyruvate.

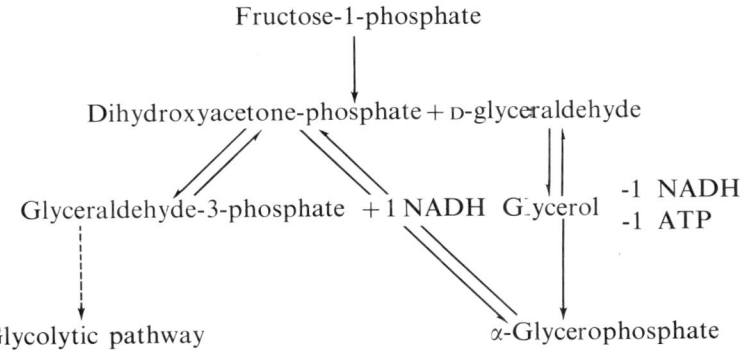

FIG. 6. Hepatic transformation of fructose-1-phosphate.

In this case, the energy balance is difficult to establish, for isomerisation equilibria can modify the distribution of the various metabolites.

The absorption rate of sorbitol, which as with all hydrogenated sugars is by passive diffusion, amounts to only 50–70% of the rate for xylitol which itself is only 50% of the rate of passive diffusion of D-glucose through the intestinal wall.[14]

According to recent studies of Förster and Mehnert,[15] the metabolic capacity of the human organism for sorbitol seems rather high, liver being able to transform 30–40 g of sorbitol/h. Such levels cannot be attained by oral administration, given that the diffusion rate of sorbitol from the intestinal lumen does not exceed 10–20 g/h, which means, in particular, that when higher doses are ingested osmotic diarrhoea may occur.

When sorbitol is introduced intravenously even at the maximum possible rate, no toxic effects can be detected. As with D-fructose, parenteral administration of sorbitol induces, during its early phase, a drop in the hepatic ATP concentration, which accounts for the fact that the newly produced D-fructose is rapidly phosphorylated. No lactic acidosis occurs.

For diabetics, sorbitol has the advantage of inducing no significant rise of blood glucose after oral administration. Though D-glucose can be produced during sorbitol metabolism, its appearance is generally delayed, so that no significant hyperglycosaemia can be detected. In diabetics, single doses of sorbitol must not exceed 10–20 g, while the total daily intake has to be limited to 60–80 g.

When sorbitol is intravenously administered, a rate of 0·5 g/kg/h must not be exceeded. Due to the absence of a kidney transport system, renal losses can reach 10% of the sorbitol administered.

METABOLISM OF D-MANNITOL

Basically, the metabolic routes of mannitol and sorbitol are similar. After passive diffusion through the intestinal lumen, D-mannitol is oxidised to D-fructose, under the action of mannitol dehydrogenase (EC 1.1.1.67) (Fig. 3):

$$\text{D-mannitol} + \text{NAD}^+ \longrightarrow \text{D-fructose} + \text{NADH}$$

Contradictory reports have been issued concerning the metabolism of D-mannitol. In fact, it seems that the negative conclusions which have been reported reflect analytical artefacts in the determination of the excreted D-mannitol.

Nasrallah and Iber,[16] having administered D-mannitol orally to human subjects in doses of 28 to 100 g, observed an absorption of more than 65% followed by a urinary excretion reaching 17·5%. After a 48-h period following ingestion, the total recovery of D-mannitol from urine and faeces amounted to 50%, the balance being oxidised to CO_2 in the liver. After intravenous injection of ^{14}C-mannitol, 18% of the initial radioactivity could be recovered as $^{14}CO_2$ in the expired air, analysed over 12 h.

When D-mannitol is given intravenously, owing to its very low rate of oxidation in liver it tends to diffuse into the extracellular region (20–25 litres for a normal subject); it then undergoes a glomerular filtration, which results in an osmotic effect with increased diuresis[17] (Fig. 7).

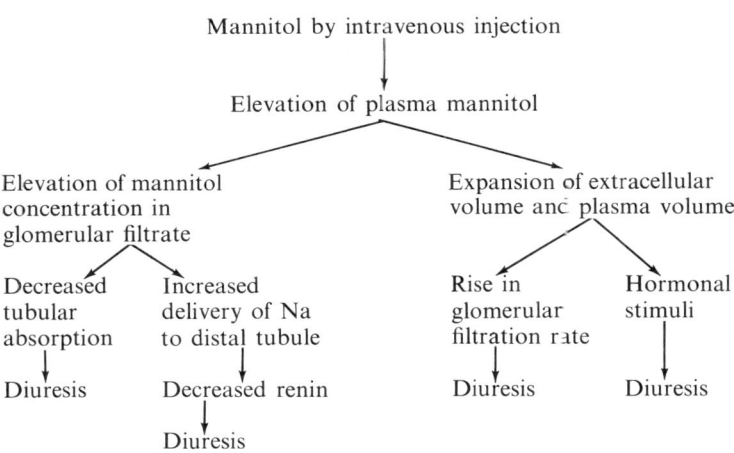

FIG. 7. Mechanism by which D-mannitol provokes diuresis.

In conclusion, the metabolic capacity of normal organisms towards D-mannitol is low and thus easy to saturate. This explains why D-mannitol has sometimes been given a caloric value of 2, due to its incomplete utilisation.

METABOLISM OF XYLITOL

In higher organisms, metabolism of exogenous xylitol generally takes place through two pathways which are closely linked: the glucuronic acid–xylulose, or Touster's, cycle, and the pentose phosphate cycle (Fig. 8).

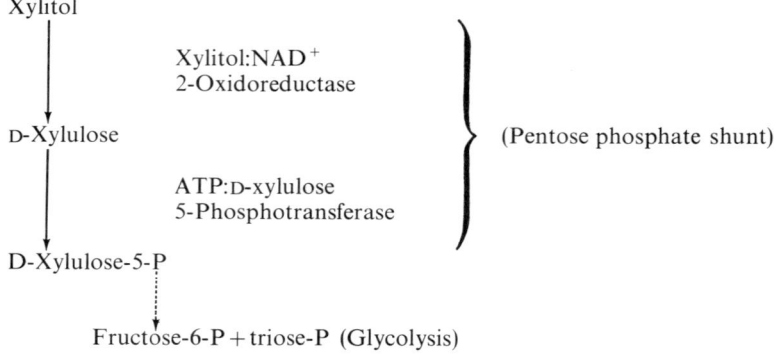

FIG. 8. Metabolism of xylitol.

The main function of Touster's cycle[18] seems to be the recycling of glucuronic acid not used for synthetic reactions or biotransformations, into glucose metabolism. In this pathway, xylitol is the intermediate in the conversion of L-xylulose to D-xylulose, which takes place in liver mitochondria.

Touster's cycle is closely connected with the pentose phosphate shunt (Fig. 9); it enters this pathway by phosphorylation of D-xylulose to D-xylulose-5-phosphate.

To enter these cycles, exogenous xylitol must first be transformed into D-xylulose, under the action of L-iditol dehydrogenase (EC 1.1.1.14), a hepatic cytoplasmic enzyme which also acts on sorbitol and similar polyols.

The energy balance of xylitol oxidation gives a theoretical total of 104 mol ATP produced when 3 mol of xylitol are consumed (Fig. 10). Thus 0·228 mol of ATP are formed when 1 g of xylitol is totally oxidised, instead of 0·211 mol of ATP in the case of D-glucose.

From the physiological point of view, the absorption of xylitol is a slow process, involving passive diffusion. This obviously limits the ingestion of xylitol, for an excess of this compound in the intestine

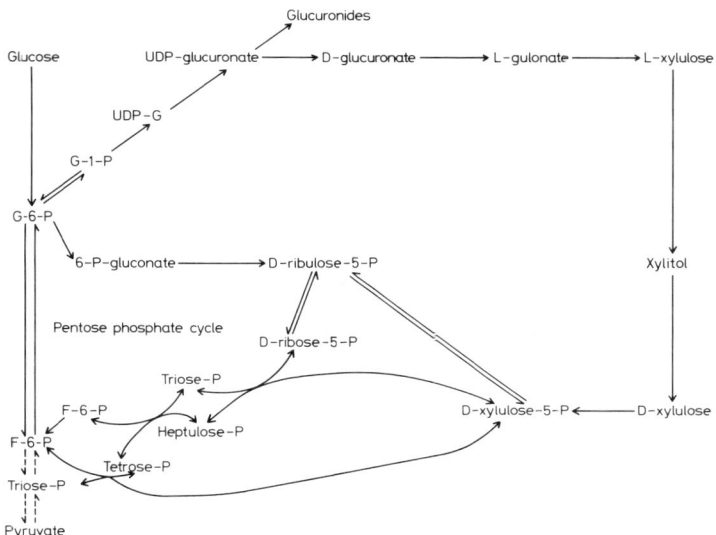

FIG. 9. Touster's cycle.

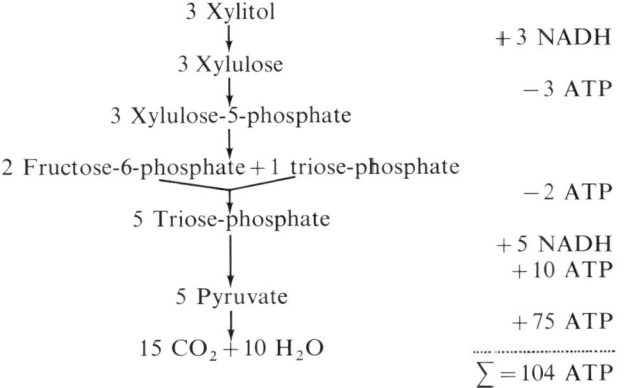

FIG. 10. Energy balance of xylitol oxidation.

can induce osmotic diarrhoea. When diffusing from the intestinal lumen, xylitol must pass across a lipoproteinic membrane. This membrane has a thickness of 10 nm and shows a porous structure, with the diameter of hydrophilic pores ranging between 0·3 and 0·6 nm.

Studies performed by Förster and Menzel,[19] using experimental methods of limited specificity have shown that the rate of intestinal absorption of xylitol is 35–60% of that of glucose despite the fact that xylitol has a molecular weight of 152 compared with 180 for D-glucose.

This unexpected difference could well be explained by conformational differences: D-glucose has a β-pyranose structure with an equatorial disposition of hydroxyl groups, whereas xylitol has an acyclic structure with hydroxyl groups on alternate sides of the carbon chain.

Though it was possible to demonstrate an adaptation in human subjects to ingested quantities of xylitol amounting to 200 g per day,[20] it seems more realistic not to exceed an overall amount of 50–70 g, spread evenly throughout the day. It must be recalled that a daily administration of 50 g/day has been used in the Turku study which lasted two years.[21,22]

Whatever its mode of administration, either orally or intravenously, about 85% of xylitol is metabolised by the liver and 10% by the kidneys; the remainder is used by various tissues. Urinary excretion remains low, even when infused intravenously.

In man, the maximum metabolic capacity lies between 0·35 and 0·50 g/h/kg body weight, equivalent to between 590 and 840 g/day for an adult man of 70 kg, when xylitol is infused intravenously. This is in sharp contrast to the capacity of Touster's cycle of 5 to 15 g/day.

In the liver, xylitol undergoes a limited transformation to glucose; the extent of this transformation which may lie between 20 and 80% depends upon the metabolic status of the organ and upon its glucose needs. This production of glucose proceeds slowly and with a delay which makes xylitol of interest as a sweetener for diabetics.

Though essentially insulin-independent, the introduction of xylitol into the energy metabolic process is not totally independent of insulin. In fact, the slow production of glucose provokes a weak stimulation of insulin release.

Due to its chemical stability, its ready metabolism and its insulin-independence, xylitol is particularly suited to parenteral nutrition of polytraumatised and diabetic patients. Furthermore low chemical reactivity allows xylitol to be used in the preparation of solutions for parenteral infusion containing free amino acids.

METABOLISM OF MALTITOL

In man, according to Dwivedi,[23] the caloric utilisation of maltitol reaches 90% of theory. When ingested, part of the maltitol is hydrolysed in the stomach, giving D-glucose and sorbitol, which then follow the normal pathways; another part of the maltitol is absorbed as such in the small intestine, to be hydrolysed by tissue enzymes.

Recently[24] it was claimed that in the digestive tract the main utilisation of maltitol involves fermentation by the gut flora to give lower fatty acids which then can be absorbed and metabolised. In fact, it is likely that the results obtained during this study were strongly dependent on the concentration of maltitol used.

Recent trials performed at Roquette[25] have demonstrated that when maltitol arrives in the small intestine, it is extensively hydrolysed to sorbitol and glucose. The absorption rate of maltitol is thus the absorption rate of its hydrolysis products, sorbitol and glucose.

HYDROGENATED SUGARS AND CARIOGENICITY

Non-cariogenicity is one of the most frequently quoted advantageous properties of hydrogenated sugars. Numerous studies have been devoted to hydrogenated sugars to establish this property. Among them the Turku study on xylitol remains a model of what can be done.[21, 22]

In Switzerland, Imfeld and Mühlemann worked on various hydrogenated sugars, using pH telemetry,[9] and the results have helped to define new types of Lycasin products and to settle the composition of Lycasin 80/55.

A rapid pre-test method was devised at Roquette which allows comparison of the rate of acidification produced by a test substance during incubation in the presence of mouth microbial flora contained in a sample of saliva (Fig. 11). It follows that if the pH remains above 6·5 after a 18-h incubation period then the product can be considered as non-cariogenic.

In fact, this pre-test is very useful as a preliminary screening of sweeteners for cariogenic potential prior to more accurate determinations.

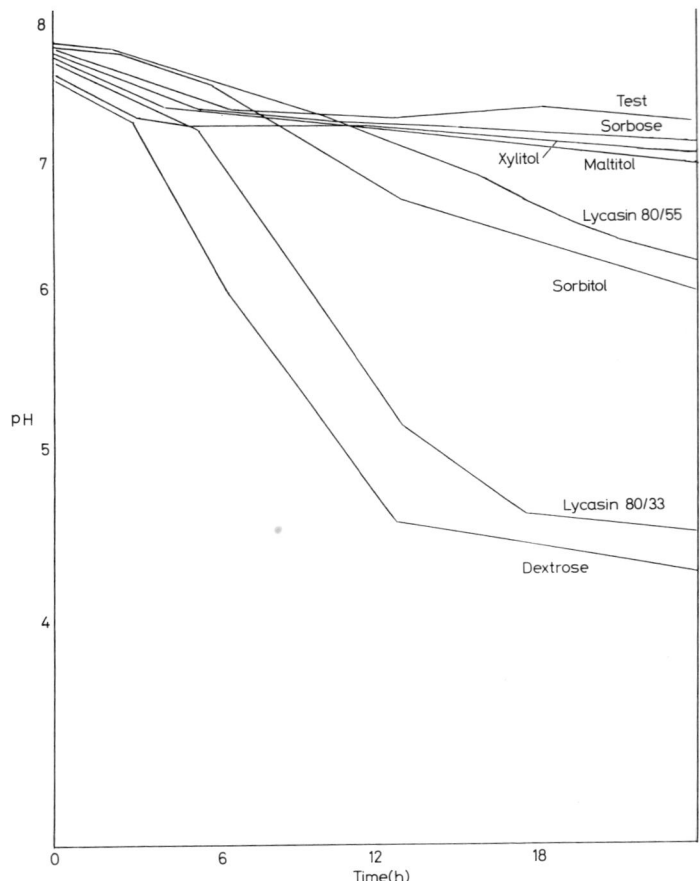

FIG. 11. *In vitro* acidification test (inoculum:saliva).

THE UTILISATION OF HYDROGENATED SUGARS IN THE FOOD INDUSTRY

It would be inappropriate to describe here, all the possible applications of hydrogenated sugars in food. Apart from their physiological behaviour, two main properties are considered when using them as replacements for sugar; their solubility in water and their relative sweetness. These two properties make possible a precise selection according to their particular application (Table 6).

In the case of mannitol, for example, the low solubility and ready recrystallisation limit the range of possible applications. Thus mannitol is mainly used as a dusting agent or as a tabletting excipient. Xylitol and crystalline sorbitol have similar applications, the former being sweeter than the latter, but much higher in price.

TABLE 6
TECHNOLOGICAL APPLICATIONS OF HYDROGENATED SUGARS

	Xylitol	Sorbitol	Mannitol	Lycasin 80/55
Hard boiled candy	±	+	±	+
Toffees, caramels	±	−	−	+
Chocolate	+	+	+	−
Chewing gum	+	+	+	+
Compressed tablets	+	+	+	−
Coatings, dragees	±	+	+	±
Bakery goods	+	+	+	+
Jellies, marmalades	+	+	−	+
Soft drinks	+	+	−	+
Canned fruits	+	+	−	+
Ice cream	+	+	−	+

Lycasin 80/55 represents a very good example of the balance between various properties. It can be used alone, as in the production of hard boiled candies, where it replaces the traditional mixture of sucrose and glucose syrup, to give a non-cariogenic confectionery. Moreover, when used in chewing-gum manufacture, Lycasin 80/55 will inhibit the crystallisation of sorbitol allowing the production of non-cariogenic gums with excellent plasticity.

The non-crystallising properties of Lycasin 80/55 make it useful in the manufacture of liquid centre filled confectionery.

CONCLUSION

It is difficult to present a coherent review of the heterogeneous data characterising the various reduced sugars. Considering their special position, in relation to sucrose, their existence seems justified by the specific applications they provide where sucrose proves unsatisfactory.

Both types of product are complementary in their properties—one single sweetener could not possibly meet all the diverse applications of sweetening agents.

REFERENCES

1. BOUSSINGAULT, (1872). *Ann. Chim. Phys.*, **26**(4), 376.
2. FISCHER, E. and STAHEL, R. (1891). *Ber.*, **24**, 538.
3. SÖRENSEN, N. A. and KRISTENSEN, K. (1950). US P. 2 516 350.
4. WHISTLER, R. L. and BEMILLER, J. N. (1973). *Industrial Gums (Polysaccharides and Their Derivatives)* 2nd Edn, Academic Press, New York.
5. *Food Chemicals Codex*, (1972), 2nd Edn, National Academy of Sciences, Washington DC, 786.
6. MANGIN, H. and HUCHETTE, M. (1972). Brev. D'Invention 7236437.
7. ROSE, R. S. and GEOPP, R. M. (1939). *The Determination of Some Physical Constants of Sorbitol and Mannitol: The Polymorphism of Sorbitol*, Atlas Powder Co.
8. CONRAD, E. (1967). US P. 3 329 507.
9. IMFELD, T. and MÜHLEMANN, H. R. (1978). *Caries Res.*, **12**, 256.
10. VERWAERDE, F., LELEU, M-M and HUCHETTE, M. (1978). Brev. D'Invention 7834830.
11. HERMAN, R. H. (1974). In: *Sugars in Nutrition*, Eds Sipple, H. L. and McNutt, K. W. Academic Press, New York, 145.
12. HERS, H. G. (1959). *Biochim, Biophys. Acta*, **37**, 127.
13. MCGILVERY, R. W. (1979). *Biochemistry: A Functional Approach*, 2nd Edn, Saunders Pub. 455.
14. FÖRSTER, H. (1978). In: *Xylitol*, Ed. Counsell, J. R. Applied Science Publishers, London, 48.
15. FÖRSTER, H. and MEHNERT, H. (1979). *Akt. Ernahrungsmedizin*, **5**, 245.
16. NASRALLAH, S. M. and IBER, F. L. (1969). *Am. J. Med. Sci.*, **258**, 80.
17. GINN, H. E. (1974). In: *Sugars in Nutrition*, Eds Sipple, H. L. and Mc Nutt, K. W. Academic Press, New York, 609.
18. TOUSTER, O. (1960). *Fed. Proc.*, **19**, 977.
19. FÖRSTER, H. and MENZEL, H. (1972). *Z.-Ernahrungswiss.*, **11**, 24.
20. DUBACH, U. C., FEINER, E. and FORGO, I. (1969). *Schweiz. Med. Wsch.*, **99**, 190.
21. SCHEININ, A. and MAKINEN, K. K. (1974). *Acta Odont. Scand.*, **32**, 383.
22. SCHEININ, A. and MAKINEN, K. K. (1975). *Acta Odont. Scand.*, **33**, (suppl) 70.
23. DWIVEDI, B. K. (1978). *Low Calorie and Special Dietary Foods*, CRC Press Inc., Cleveland, Ohio, 19.
24. RENNHARD, H. H. and BIANCHINE, J. R. (1976). *J. Agric. Food Chem.*, **24**, 287.
25. ROQUETTE FRERES, Unpublished observations.

9

Nutritional Significance of Sweetness

M. NAIM
The Hebrew University of Jerusalem,
The Levi Eshkol School of Agriculture,
Faculty of Agriculture, Rehovot 76-100,
Israel

and

M. R. KARE
Monell Chemical Senses Center,
Philadelphia,
Pennsylvania, USA

ABSTRACT

Experimental data suggest that the responses to sweetness in both humans and animals are innate. For most mammals, a sweet food is palatable (historically more nutritious) and would usually be strongly preferred to bitter foods (aversive and poisonous, in many cases). These innate responses and learning processes which make use of taste as a cue, can serve in choice situations as a protector against consumption of foods of questionable quality. In a free food choice situation, experimental animals can use the above mechanism in order to select nutrients according to physiological needs.

Food for humans, however, is obviously prepared by another human and therefore, socio-cultural factors predominate in food choices. On the other hand, the physiological mechanism by which oral factors can trigger eating in man does not seem to be so different from animals since exposing animals to various external sensorial stimuli can lead to hyperphagia. For example, giving diets containing an appealing taste stimulus to rats does not usually stimulate overeating, but offering a 'cafeteria set-up' of various palatable foods will produce dietary obesity.

While the role of taste in food selection is obvious, less attention has been given to the effects of oral stimulation on the systemic physiology of the organism. However, the oral–cephalic relay provides information to the central nervous system, which responds by initiating appropriate endocrine and exocrine secretions. These include the stimulation of salivary and gastric acid secretion, the release of gastrin, an increase in pancreatic exocrine output and mobilisation of insulin from the endocrine pancreas. These activities may involve motility of the gastro-intestinal tract, and are, at least partially, autonomic in their neural mediation.

The question arises as to whether or not these gustatory–digestive and metabolic pathways have nutritional significance. Most of the responses described above can be initiated without the orogastric reflexes.

Experiments with dogs have demonstrated that taste stimulation alone does not affect pancreatic exocrine output. However, when coupled with swallowing, there is a greater effect by palatable (sweet) than by unpalatable (bitter and sour) taste stimuli on the cephalic phase of pancreatic exocrine secretion. Furthermore, experiments with rats indicated that the pre-absorptive insulin response to palatability is positive for sweet taste but absent following salt stimulation. The present discussion will focus attention on the possible role of the above phenomena in nutrition.

INTRODUCTION

Classic studies have explored the physiological role and mechanisms of the action of specific nutrients. Reliable guides of dietary recommendations delineate the minimum amounts of essential micro and macro nutrients that diets should contain. International feeding programmes have directed the attention of nutritionists to the sensory aspects of hunger, satiety and diet selection.

One important nutritional role of the sensory mechanism at the oral level is to initiate eating and to assure a reliable quality control of nutrient intake. Lepkovsky[1] stated that the 'universality of the chemical senses is best explained by the fact that in order to eat, the organism must possess some chemical knowledge'. Some mammals have a 'sweet tooth' as evidenced by their selection of sweet-tasting foods. Because certain simple sugars are appealing to some species, the evolutionary role of sweetness in energy homeostasis is suggested. It should be noted that many simple carbohydrates evoke no response in many species.

Taste should not be considered solely as an external cue[2] since its perception and its role in eating are heavily dependent upon the physiological state.[3-6] The need for nutrients initiates metabolic activities which are transferred to the central nervous system where they can be expressed as appetite. Conversely, oral sensation may initiate some endocrine and exocrine secretions of digestion and metabolism[7-10] and therefore may affect physiological activities.

The significance of sweetness in nutrition will be evaluated in this paper. The function of sweetness will be considered in relation to palatability, eating behaviour, digestion and metabolism.

THE ROLE OF ORAL FACTORS IN FOOD CHOICES

In humans, the response to sweetness is probably innate. Newborns prefer sugar solutions to water and their preference increases at higher concentrations.[11] Further, their preference for concentrations of, as well as for different, sugars corresponds with adult human perception.[12] In contrast, intake data for salt and bitter solutions at the concentration tested did not show either preference or aversion in newborns.[13] Studies examining the facial expressions of neonates following taste stimulation confirmed the responses to sweetness, and also suggested that infants find quinine at high concentrations aversive.[14] In rats, taste buds appear to reach full maturity 10-12 days postnatally.[15] This may explain why neonatal rats do not respond to sucrose and saccharin until about the 7th day postpartum.[16] Using the facial examining technique, it is possible to discern that neonatal rabbits at ages of less than 24 h postpartum are able to respond differently to sucrose and quinine.[17]

When given a choice of novel foods, rats will initially select diets of appealing flavour rather than diets containing aversive compounds. This discrimination without apparent learning might have had survival value in the phylogenetic history of mammals and therefore was genetically retained. For example, many appealing (including sweet) flavours are associated with nutritious foods, aversive flavours with poisons.

Animals generally display a slow acceptance of novel foods; this assures them of enough time to make quality assessment.[18] More common are the mechanisms wherein the animals associate the sensory quality of the food with postingestional consequences. Postingestional factors refer to those effects occurring after materials are ingested, including gastro-intestinal tract processes and postabsorptive metabolic effects.[19-21] In making behavioural adjustments to regulate their internal environments, animals will quickly learn to avoid substances that are followed by toxic aftereffects. When associated with aversive events, previously neutral or preferred stimuli become aversive themselves.[22] This type of learning has been termed 'conditioned taste aversion'.

Even though responses and learning processes are important in developing taste preference, preferences may be altered with the physiological state. For example, human subjects prefer 5% sucrose solution to 30% sucrose solution, but can be induced to change this preference after blood glucose levels are decreased to 50 mg% following insulin injection.[23] Similar results were later found in rats.[5] Insulin administration produces an almost immediate hunger for sweetness.[6] Likewise, adrenalectomised rats will develop a specific hunger for salt, and sodium-deficient rats display an immediate recognition of sodium from among other cations.[24, 25] It appears that the specific hungers described for sweetness and salt are, at least, partly innate since almost no learning time was required for the animals to show the specific behaviour. A much longer time is needed, on the other hand, for rats deficient in B vitamins to select the diet containing the vitamins from among several choices.[3, 26] Similar learning processes were probably operative in an experiment the results of which are presented in Fig. 1. Young rats were given a choice between two diets differing not only in flavour but also in nutritional potential.[27] One diet contained defatted raw soybeans as a protein source with the addition of 0·35% (w/w) sodium saccharin which has an appealing taste at this level. The other diet contained defatted heated soybeans with the addition of the aversive stimulus, 2·0% (w/w) sucrose octaacetate. Raw soybeans have a lower nutritive value than the cooked soybeans because they contain digestive enzyme inhibitors and other undesirable thermolabile factors.[28]

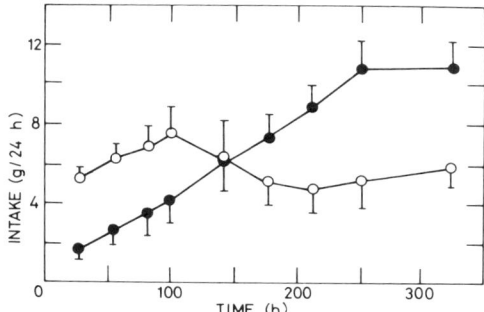

FIG. 1. Preference test between a diet containing raw soybean flakes mixed with 0·35% sodium saccharin (○) versus a diet containing heated soybean meal mixed with 2·0% sucrose octaacetate (●). Values are average food intake in grammes and SEM of 12 rats. Adapted from Reference 27.

Given a choice between these two diets, the rats initially preferred the diet with the better taste (poor nutrition). After 6 to 7 days, the rats changed their preference to the diet with the better nutrition even though it contained the aversive taste of sucrose octaacetate. This change in preference may be explained by assuming that postingestional factors eventually influenced the animals to select the diet offering better nutrition. It therefore appears that in a choice situation, animals under some circumstances are able to select nutrients according to their nutritional needs.

Food for humans is commonly prepared by other humans. Since religious and social factors contribute to the development of food preferences, consumption patterns are more likely to be influenced by variables independent of flavour.[18] Food choices are determined to a large extent by availability and by the cuisine of the particular culture. Since cuisine usually has a nutritional advantage, humans are generally exposed to safe foods after cooking.

Taste preference in man can be related to previous intake experience.[29] Moskowitz et al.[30] found that labourers in India, whose diet contains many sour foods, rated citric acid as pleasant with increasing concentration. On the other hand, Indian medical students, who usually eat Western cuisine, rated citric acid as aversive in a pattern similar to that found in the Western population.[31] Furthermore, exposing rats to citric acid during weaning showed an increased voluntary ingestion of citric acid as compared to animals exposed to citric acid after weaning or not exposed to citric acid at all.[32] Also in rats, tastes transmitted through mothers' milk may affect food preferences in the post-weaning period.[33] Basic mechanisms of diet selection in humans and laboratory animals may be similar, but this fact may not be readily observed since laboratory animals, unlike humans, are not normally exposed to a variety of flavours during their physical development.

The significance of sweetness in food choices is of importance because of its possible role in both normal and clinical nutrition. Obese subjects show 'hyper-responsiveness' to high levels of palatability. For example, in one study,[34] overweight subjects ate significantly more good-tasting food than subjects of normal weight. Rodin[35] found that overweight subjects rate higher concentrations of glucose as more pleasant than normal subjects, but Grinker et al.[36] indicated that obese subjects find higher concentrations of sucrose aversive. The later finding contradicts the hypothesis that abnormal sweetness preference in an obese subject is a contributing factor to obesity. Although the results of different studies are inconsistent, it appears that the perception of sweetness *per se* does not contribute directly to obesity. However, the overweight subjects do ingest more of what they like.[35] It is possible, therefore, that overingestion by the obese could be due to a general hyper-responsiveness to all foods rather than to sweetness alone. One should not be surprised, however, to find a greater variance in obese subjects' responses derived from different studies. Studies of various forms of obesity, i.e. dietary obesity and different syndromes of genetic obesity,[37, 38] have revealed that some forms may be developed primarily due to hyperphagia, whereas others are caused by metabolic abnormalities. The obesity syndrome and its associated abnormalities, such as hyperinsulinaemia and high levels of corticosteroids, can be prevented by caloric restriction in some models, whereas, in others, the syndrome is not corrected by food restriction. Thus, the metabolism and the mechanisms triggering eating may vary markedly from one obese subject to another.

A significant impairment of sweet-taste detection has been reported in diabetic patients.[39, 40] Experimentally induced diabetic rats also demonstrated behavioural changes in their preference for sweet solutions.[41] In humans, adult-onset diabetics and their first-degree relatives showed a significantly higher glucose threshold than control subjects.[40] The adult-onset diabetics, but not their relatives, also demonstrated a higher sucrose threshold as compared to controls. One hypothesis suggested that the taste alterations in diabetics are due to a 'satiation effect' produced by the elevated blood glucose levels.[42] Another hypothesis is that the decreased sensitivity is due to a generalised defect in cellular glucose sensitivity in both the pancreatic beta cells and in the taste cells.[43] The decrease in sweet-taste sensitivity was thought to cause increased sugar ingestion in diabetic patients in order to produce the pleasant sensation. Preference tests, however, indicated that the decreased sensitivity was not related to sweetness preference.[40] More longitudinal studies are needed in order to evaluate the significance of possibly elevated sweetness thresholds in diabetics.

PALATABILITY AND FOOD CONSUMPTION

Central and peripheral mechanisms of food intake have been a focus of increased attention.[2, 44–46] One would expect that oral factors, which contribute significantly to the regulation of diet selection, would also determine the food-seeking behaviour and the amount of food eaten. However, studies with experimental animals suggest that the quantitative regulation of food intake can be accomplished without the oropharyngeal factors. Rats, for example, may be trained to feed themselves by pressing a lever that delivers liquid food into the stomach.[20] Furthermore, voluntary intragastric feeding in man indicated that a single meal can be identical to oral feeding with respect to the rate of intake and the amount consumed.[47] Oral factors can, however, contribute to satiety in rats[19] and dogs.[48] In sham feeding experiments, in which food is diverted to the exterior before entering the stomach, voluntary oral feeding eventually ceases, but this cessation takes longer compared to normal feeding. Human subjects, who simultaneously ingest food through the mouth and through a gastric tube, will eat excess amounts of food,[47] thus suggesting that both oral factors and postingestional signals are needed for the optimal regulation of eating.

The magnitude of the effect of food palatability on consumption is still undefined. Several studies suggest that neither growing rats[49, 50] nor adults[51] will reduce their food intake when eating a diet adulterated with an aversive taste stimulus. The reduced food intake and weight loss produced by a quinine-adulterated diet[52] may not be due to sensory properties alone.[53] Sensory adaptation or habituation to the aversive taste stimulus may develop, allowing the animals to consume nutrients according to physiological needs. These results suggest that postingestional signals are of greater consequence to the rat than the flavour of the diet; however, habituation may mask the taste effects. In an attempt to interfere with such protecting mechanisms, growing rats were recently[54] fed a 10% protein (casein) diet the taste of which was changed daily by the addition of one of four aversive taste stimuli in a pre-determined order. The stimuli added to the diet (w/w) were: 3% sucrose octaacetate, 2·5% sodium saccharin, 4·0% sodium chloride and 3·0% citric acid. These stimuli are aversive to rats at these levels in a choice situation. Weanling male rats were divided into three groups, each containing 13–15 animals (Fig. 2).

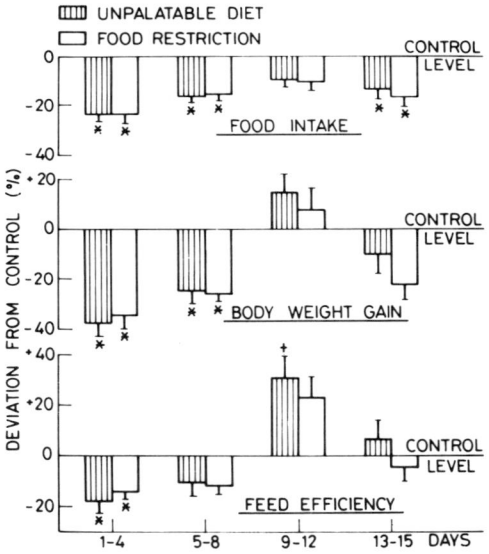

FIG. 2. Non-cumulative values of food intake, body weight gain and feed efficiency within each period of 3–4 days, for rats fed the unpalatable diet and for their pair-fed animals fed less than *ad libitum* amounts of the unadulterated diet. Bars show the group means (and SEM) as percentage deviation from the control group fed the unadulterated diet *ad libitum*. *Significantly lower value (at least $p < 0.05$) than the control level. †Significantly higher value ($p < 0.05$) than the control level. From Reference 54.

The first group (control) was fed *ad libitum* an unadulterated casein diet. The second group was fed *ad libitum* the casein diet that was modified by a daily addition of one of the above aversive stimuli (unpalatable diet). Rats of the third group (food restriction) were pair-fed each day an amount of unadulterated casein diet according to the consumption observed by their weight-matched animals in the second group during the previous day. During the 15-day experiment, rats fed the unpalatable diet ate 10–20% less food compared to controls fed the unadulterated diet. The unpalatable diet also induced changes in body-weight gain and feed efficiency (body-weight gain/food intake), but similar effects were found in the third group of rats fed restricted amounts of the unadulterated diet. The magnitude of these changes in body weight will be discussed in the following section on digestion. It was suggested that daily aversive taste change is an effective way to reduce food intake in growing rats. The reduction in food intake did not, however, exceed the average level of 15%. Undoubtedly, the need for energy in these young animals was a much stronger signal in determining food intake than the aversive taste of the diet.

Could highly palatable diets increase food intake? In rats, dietary 'lab chow' is usually taken as a baseline of normal food intake and body weight gain. When appealing levels of sodium saccharin (0·25–0·35%) are supplemented to the above stock diet or given as a separate solution in addition to water, the *ad libitum* caloric intake level of normal non-obese rats was not affected.[55] Internal signals seem to be incapable of guiding a stable caloric intake in humans,[56,57] yet studies have shown that experimental animals make appropriate adjustments to maintain caloric balance. One would, therefore, assume that the use of animals as a model for human eating behaviour is unjustified. However, the sensory properties of the food alone can produce a conditioned satiety in laboratory rats.[58,59] LeMagnen[58] used 'odour-labelled diets' to show a specific satiation effect of food stimuli at the peripheral level and its acquisition by learning (Fig. 3). During a 32-day period, rats received a daily 2-h test meal. On successive days, the diet presented was flavoured by any one of the added odour substances: citral, eucalyptol, benzyl acetate or benzaldehyde (A, B, C and D in Fig. 3).

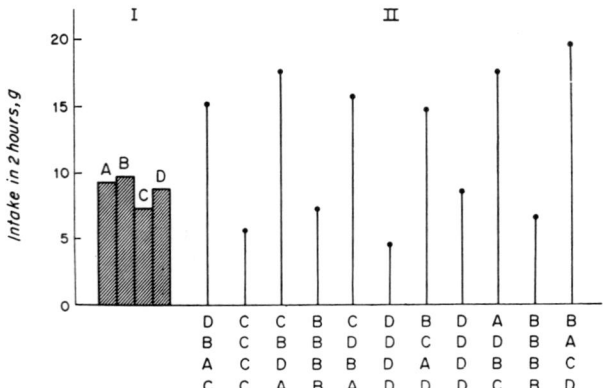

FIG. 3. I. Average food intake during the control period. II. Average food intake in meals with either successive or constant flavoured samples. From Reference 58.

At the end of the 32-day period, the amounts of food eaten were found to be the same for all four types of odour-flavoured diets. Then, during the same type of test meal, the four different diets were presented successively for 30 min each. The tests were performed at 48-h intervals. On days in between, the meal contained a single odour-flavoured diet during the two hours as indicated in Fig. 3. Rats fed from the first sample during the 30 min ate an almost equal amount as the normal control level that was previously consumed in the 2-h meal. Then, from the subsequent sample, they ate again as if they had not already eaten. Under these circumstances, their average intake was 70% higher than in the control meal. Thus, hyperphagia could be provoked by external sensory stimuli. The observed satiety is therefore different from the normal satiety which limits consumption according to metabolic needs. Similarly, Sclafani and Springer[60] found that feeding adult rats a variety of snack foods along with their dietary chow is an effective way of producing dietary obesity. In addition to the lab chow and water, female rats were given a high fat (33%) diet and a food assortment, such as biscuits, salami, cheese, bananas, peanut butter and others. Body-weight gain of the experimental groups was over 2·6 times higher than in the control group fed lab chow only. The experimental animals did not maintain their obesity when given only lab chow, but rather reduced their weight to control level. Food intake measurements were not systematically taken, but the investigators indicated that the experimental group ate much of the snack foods and little of the lab chow. Since dietary components (e.g. high fat) may increase the feed efficiency and induce obesity even without hyperphagia,[61,62] further research is needed to quantify more precisely just how much the palatability contributed to the above phenomenon. These studies however, demonstrated that rats as well as humans can become hyperphagic when exposed to various motivating stimuli. It is not the palatability of the diet *per se* (i.e. how sweet or salty the food is) which made the animals hyperphagic, but rather the anticipation of satisfying various qualities of sensation which are conditioned by learning. It may be concluded that relatively drastic modifications in diet palatability are required in order to reduce food consumption (Fig. 2) or increase it (Fig. 3) over the long term.

EFFECT OF ORAL STIMULATION IN DIGESTION AND METABOLISM

Oral stimulation can modify the secretory process along the gastrointestinal tract. Pavlov and his colleagues[63] were the first investigators to show that chewing of palatable substances by dogs stimulated gastric secretion and that little secretion was noted for neutral-tasting substances. Stimulation of digestive juice secretion can be classified into three phases: cephalic, gastric and intestinal.[64] The cephalic phase is operated by several stimuli including anticipation, taste, smell and sight of food. In the conscious state, the cephalic phase can be isolated by sham feeding in which ingested food is diverted to the exterior, usually by an oesophageal fistula. The cephalic phase of gastric secretion is mediated entirely by the vagus since intact vagi are required for cephalic phase induced reflexes during sham feeding.[63] The gastric secretory response to sham feeding appears in dogs after a latent period of 5–7 min and may continue as long as 3 h after the food is eaten. The response to sham feeding was also demonstrated in man.[65] In humans, it has been suggested that when all phases (cephalic, gastric and intestinal) are operating simultaneously, the cephalic phase is responsible for one-third of the total amount of gastric acid secretion.[10]

Stimulation of pancreatic exocrine secretion also occurs in cephalic, gastric and intestinal phases. While the cephalic phase of gastric secretion is known to play an important physiological role, the pancreatic cephalic phase is regarded as being of little consequence. However, sham feeding in man and dogs does produce pancreatic exocrine secretion.[8,66] In dogs, a portion of the pancreatic response during sham feeding is due to vagally induced gastrin released from the pyloric antrum. Pancreatic exocrine response during sham feeding was markedly reduced by stomach acidification which is known to inhibit the vagal release of gastrin.[66] In man, vagally released gastrin appears to be of less importance. Plasma gastrin levels were not significantly increased during sham feeding in duodenal ulcer patients[67] or in anticipated feeding in healthy men.[68] The pancreas has vagal innervation and there is evidence of direct neural stimulation of pancreatic secretion.[69] It appears that, in man, this mechanism is more important than the vagally mediated gastrin response. Since gastric acid secretion is stimulated during the cephalic phase, it further increases pancreatic volume and bicarbonate output by releasing the duodenal hormone secretion.

While stimulation of receptors at the oral level can affect digestive juice secretions, limited consideration has been given to the role of taste in these exocrine secretions. Experiments with conscious dogs have indicated that both pancreatic flow and pancreatic protein output can be affected by the nature of the taste stimuli.[70] A subsequent study[49,71] was designed to explore further the relationships of specific taste stimuli to pancreatic secretion. Dogs were prepared with gastric and duodenal fistulas,[72] thus allowing a continuous drainage of gastric and intestinal contents during experiments as well as exposing the main pancreatic duct for direct cannulation.[73] Stimulation of gustatory receptors was by application with swabs of aqueous solutions to the dogs' tongues for 6 min. Sucrose, in a concentration to produce palatable taste, and quinine sulfate or citric acid in concentrations to produce unpalatable taste were selected. These stimuli were presented in a pre-determined experimental design and saliva was used as a control stimulation. Not more than one stimulation a day and three stimulations a week per dog were applied. During experiments, gastric and duodenal fistulas were kept open.

Nutritional Significance of Sweetness

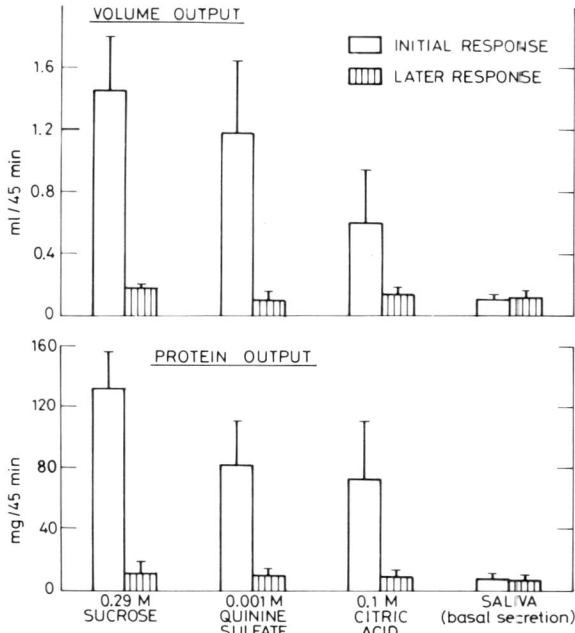

FIG. 4. Total pancreatic volume and protein output during 45 min following oral stimulation with three different taste stimuli using the swab technique. Values are the mean and SEM of four experiments for each stimulus during either the initial or later sessions using four dogs. Adapted from Reference 49.

At first, oral stimulation with these substances produced a large increase in both the pancreatic volume and protein output during the 45 min of the collection (Fig. 4). However, after one or two trials with each stimulus with each dog, the increased output was no longer seen. It is probable that the dogs associated the initial taste stimulations with anticipated feedings; however, when they learned that food would not be forthcoming, they no longer responded. The phenomenon of reduced response (extinction) following repeated trials has not been reported with sham feeding, but the orogastric reflexes, besides being excited by gustatory impulses, are subjected to inhibition by the hypothalamic area.[74] Since swallowing is usually involved with sham feeding, the same dogs were next given, orally, 100 ml of taste stimulus solution mixed with 25 g of cellulose (Fig. 5).

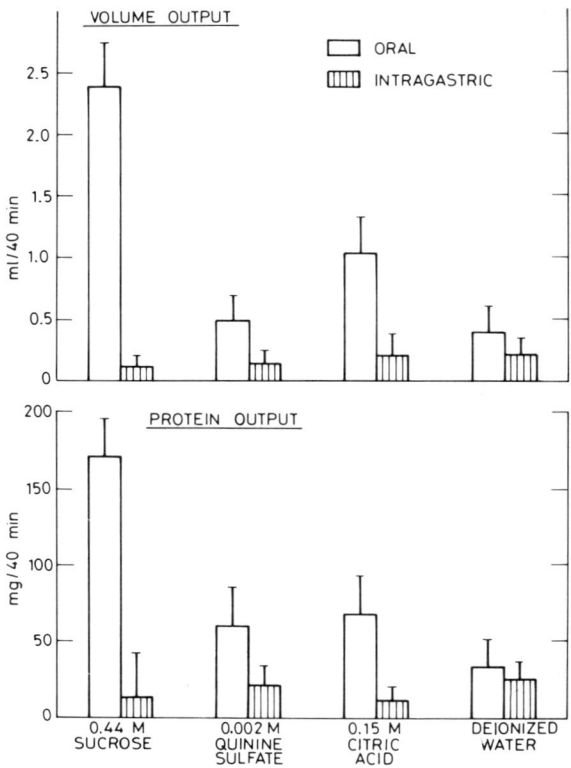

FIG. 5. Total pancreatic volume and protein output during the 40 min following oral or intragastric administration of taste stimulus – cellulose mixtures. Values are the mean and SEM of 6–8 experiments using four dogs. Adapted from Reference 71.

The pancreatic secretory response which occurred within 40 min following the 6-min period of oral administration was restored primarily for a sucrose–cellulose mixture. Oral administration of the unpalatable citric acid–cellulose or quinine–cellulose mixtures resulted in low pancreatic output, similar to that for the control water–cellulose mixture. Although the gastric and intestinal fistulas were kept open, intragastric administrations of the same taste stimulus–cellulose mixtures were performed in order to separate any pre-gastric factors that affect pancreatic secretion from those due to postingestinal stimulation, e.g. the vago-vagal reflex.[75] These intragastric experiments resulted in low pancreatic output during the 40 min after administration. Thus, the swallowing action was necessary for triggering the feedback mechanisms which could restore the pancreatic response.

Schwartz et al.[76] also found that, in man, the cephalic phase of pancreatic polypeptide secretion from the endocrine pancreas was larger and less variable when the food was not only tasted and chewed, but also swallowed. We concluded that taste stimulation alone is not sufficient to affect pancreatic exocrine secretion, but that when coupled with swallowing, palatable taste stimuli exert a greater effect than do unpalatable stimuli on the cephalic phase of pancreatic exocrine secretion. Using a similar design, Ohara et al.[77] reported recently that the effect of gustatory stimulation on pancreatic exocrine secretion of Beagle dogs was dependent on the carrier used. Statistically significant increases in pancreatic response to an oral sucrose–cellulose mixture, as compared to water–cellulose stimulation, were not found in these dogs. However, the mean for sucrose stimulation was about 18 times higher than that for water. It is apparent that the individual responses varied markedly (high standard error level) among these dogs and this could explain their results.

As most of the digestive enzyme content in the gastrointestinal tract is secreted from the exocrine pancreas, one should explore whether the cephalic phase of pancreatic secretion can affect food utilisation. Early digestive juice secretions which occur due to oral stimulation are followed later by the same qualitative pattern of secretions in response of postingestional stimulation. Thus, the important role of the early cephalic exocrine responses may indeed be early preparation for food digestion. The gastric and the intestinal phases are known to account for the larger part of pancreatic exocrine secretion. Smith et al.[78] (cited from Lepkovsky[1]) served unpalatable foods to human subjects. They indicated that the food was so unpalatable that one subject vomited shortly after eating. However, only a small difference in food digestibility as measured by nitrogen retention was found between subjects fed the unpalatable diet and controls fed palatable food.

Similar results were found with rats in our recent study (see Fig. 2). When we fed weanling rats a 10% protein diet containing an aversive taste stimulus whose character was changed daily, the feed efficiency (body-weight gain/food intake) was reduced during the first 8 days as compared to controls fed the unadulterated diet. However, the results have demonstrated that the reduced food intake alone (restricted amounts of unadulterated diet given to the pair-fed animals) inhibited feed efficiency to the same extent. Furthermore, an additional experiment of nitrogen balance showed that faecal nitrogen was not increased in rats fed the unpalatable diet as compared to controls fed the unadulterated diet. This ruled out the hypothesis that the digestive efficiency was reduced. Rather, a higher level of urinary nitrogen was found in rats fed the unpalatable diet. Thus, part of the ingested protein was metabolised for energy rather than for anabolic processes in order to compensate for a caloric deficit which occurred due to the reduced food intake. As shown in Fig. 2, the feed efficiency of rats fed the unpalatable diet was 30% above the control level during days 9–12. The result suggests that the initial inhibition of feed efficiency observed during days 1–8 is compensated for in days 9–12 so that overall feed efficiency for the entire 15-day period is identical for experimental and control subjects. It was concluded that the feeding of the unpalatable diet as well as imposing a 15% food restriction could affect the feed efficiency for only short periods.

For several reasons, we suggest that the effect of diet palatability on food utilisation is still only partially understood and should be further determined. First, the feed efficiency experiments should be carried out for longer than a two-week period. At the end of such experiments, body composition should be determined. Secondly, our preliminary unpublished results and other data[52] indicate that the meal pattern may be affected by diet palatability. If so, the meal pattern changes may affect the way in which nutrients are used.[79] Thirdly, our additional results[80] demonstrated that the proteolytic digestive enzymes content in the faeces is markedly higher in rats fed the unpalatable diet. On-going experiments are designed to explore this phenomenon.

The normal over-optimum level of digestive enzymes occurring in the small intestine may explain the lack of effect of the unpalatable diet on digestibility. One may hypothesise that such over-optimisation of digestive enzyme does not occur under specific circumstances, e.g. in the clinic. Wolf and Wolff[81] reported that their patient, Tom, who had to be tube-fed directly to the stomach was 'poorly nourished' and that such feeding 'failed to satisfy his appetite'. When Tom, at his own suggestion, began to taste the food, however, he gained weight and showed a 'good appetite'. Similarly, Eddie, the patient of Hollander et al.,[82] showed a 'persistence in his desire to eat ordinary foods' even though they were soon 'regurgitating through the oesophageal sac'. Presumably, the activation of receptors in the mouth promoted better nutrition.

The neurohumoral mechanisms which control pancreatic exocrine secretions may also initiate pancreatic endocrine secretions.[83] The oral delivery of glucose into the intestine in man is followed by the development of higher blood levels of immunoreactive insulin (IRI) than those obtained when the same amount of glucose is administered intravenously.[84, 85] In dogs, sham feeding of glucose or tap water alone could increase the IRI.[9] This phenomenon was, however, absent after topical anaesthesia of the oral mucosa,[86] suggesting that the initial release of insulin is neurally mediated upon stimulation of receptors in the mouth. In rats, the initial IRI peak after feeding is entirely dependent upon oral stimulation and occurs within one minute of the stimulation.[87] Nicolaidis[88] found a slight hyperglycaemia after either oral sucrose or oral saccharin stimulation in anaesthetised food-deprived rats. He has hypothesised[89] that during food deprivation, the anticipatory processes of releasing endogenous glucose prevail, resulting in early hyperglycaemia, while during *ad libitum* feeding, glucose storage processes are predominant. This suggestion is in line with the finding that the release of insulin during the first minute of feeding is much smaller in food-deprived rats than in rats fed *ad libitum*.[90] Furthermore, glucagon is also released in rats within 1–2 min following feeding.[91] In rats, gustatory stimulation by glucose and saccharin though not by sodium chloride could elicit the preabsorptive insulin release even though they were all presented at appealing levels.[92, 93] This suggests that the behavioural responses can be dissociated from the autonomic responses which elicit the insulin release. Based on results of various studies[86-88], one may suggest that the preabsorptive insulin release in response to sweet taste stimulation is an innate reflex.

After conditioning, increased insulin level can be obtained by other stimuli.[94, 95] A rise in plasma insulin levels during imaginary food ingestion under hypnosis of voluntary men has been reported.[96] As for pancreatic exocrine function, almost all of these endocrine secretions, in response to the cephalic phase stimulation, are observed in later stages in response to postabsorptive effects.

Atropin injections[97] or vagotomy[98] can usually block the cephalic insulin response.

It is still questionable whether the insulin release after sweet taste stimulation is an innate response. Deutsch[99] found that the hypoglycaemic effect induced by saccharin ingestion was extinguished in rats by exposing them to long-term access to saccharin. More recently, in a series of experiments, Berridge et al.[93] have demonstrated that the preabsorptive release of insulin evoked by preferred taste stimulus (i.e. glucose) can be abolished by pairing that taste with LiCl-induced illness. Their results may indicate that experience could change the neuroendocrine function and the ingestive consummatory behaviour as well.

Could these early neuroendocrine reflexes have any physiological or nutritional significance? As for the digestive secretions, the postabsorptive effects have a larger quantitative portion than preabsorptive stimuli in the overall secretion response. According to Nicolaidis,[89] the sensory endocrine reflexes have an optimising role in the metabolic use of nutrients. Oral ingestion of saccharin, for example, improved the anabolic utilisation of nutrients administered intravenously. The more moderate increase in blood glucose level observed in normal oral feeding as compared to intragastric feeding[87] may also display an optimising role of the pre-absorptive insulin release. The potentiation by saccharin ingestion (presumably due to its sweet taste) of the insulin-induced coma[99,100] also suggested a possible physiological significance for the preabsorptive insulin release. A more relevant role in nutrition and physiology of the cephalically induced insulin release, might be found, however, in its promotive function in the initiation of eating. Prolonged insulin administration can lead to increased food intake[101] and to changes in food choices.[5] Conversely, hyperinsulinaemia is a common characteristic found in various forms of obesity.[37,38] Since taste, smell and sight of food can induce the early insulin release, one may assume that such mechanisms will contribute to subsequent motivation towards eating. This hypothesis has been well-described by Rodin[2] who found that individuals whose eating behaviour was most responsive to external cues associated with food, showed the greatest insulin release in response to the sight and smell of grilling steaks. In a recent study, Sjöström et al.[97] found that the insulin level was not elevated in normal weight women in response to food presentation, but was significantly elevated in obese subjects. These investigators suggested that the insulin response might, to a certain degree, be an index of appetite. Thus, the hyper-behavioural responses to food described for some obese subjects[34, 35] could be partly related to the cephalically induced insulin release.

ACKNOWLEDGEMENTS

The authors wish to thank Dr Harvey J. Grill and his colleagues for permission to read their pre-published data, and Dr Joseph G. Brand for his assistance in the preparation of this manuscript.

REFERENCES

1. LEPKOVSKY, S. (1977). In: *The Chemical Senses and Nutrition*, Eds Kare, M. R. and Maller, O. Academic Press, New York, 413–28.
2. RODIN, J. (1978). In: *Recent Advances in Obesity Research*, Vol. 2. Ed. Bray, G. Newman, London, 75–85.
3. HARRIS, L. J., CLAY, J., HARGREAVES, F. and WARD A. (1933). *Proc. Roy. Soc. Lond. (Biol.)*, **113**, 161–90.
4. RICHTER, C. P. (1936). *Am. J. Physiol.*, **115**, 155–61.
5. JACOBS, H. L. (1958). *J. Comp. Physiol. Psychol.*, **51**, 304–10.
6. DAVIS, J. D. and LEVINE, M. W. (1977) *Psychol. Rev.*, **84**, 379–412.
7. PETHEIN, M. and SCHOFIELD, B. (1959). *J. Physiol. (Lond.)*, **148**, 291–305.
8. NOVIS, B. H., BANKS, S. and MARKS, I. N. (1971). *Scan. J. Gastroenterol.*, **6**, 417–21.
9. HOMMEL, H., FISCHER, U., RETZLAFF, K. and KNÖFLER, H. (1972). *Diabetologia*, **8**, 111–6.
10. RICHARDSON, C. T., WALSH, J. H., COOPER, K. A., FELDMAN, M. and FORDTRAN, J. S. (1977). *J. Clin. Invest.*, **60**, 435–41.
11. DESOR, J. A., MALLER, O. and TURNER, R. (1973). *J. Comp. Physiol. Psychol.*, **84**, 496–501.
12. MALLER, O. and DESOR, J. A. (1974). In: *Oral Sensation and Perception: Development in the Fetus and Infant*. Ed. Bosma, J. US Govt. Printing Office, Washington, DC, 279–91.
13. DESOR, J. A., MALLER, O. and ANDREWS, K. (1975). *J. Comp. Physiol. Psychol.*, **89**, 966–70.
14. STEINER, J. E. (1973). In: *Fourth Symposium on Oral Sensation and Perception*, Vol. 2, Ed. Bosma, J. F. US Dept. of H.E.W., Bethesda, Md., 54–258.
15. MISTRETTA, C. M. (1972). In: *The Third Symposium of Oral Sensation and Perception: The Mouth of the Infants*, Ed. Bosma, J. F. Thomas, Springfield, Ill., 163–87.

16. JACOBS, H. L., SMUTZ, E. R. and DUBOSE, C. N. (1977). In: *Taste and Development: The Genesis of Sweet Taste Preference*, Ed. Weiffenbach, J. M. US Dept. of H.E.W., Bethesda, Md., 99–108.
17. GANCHROW, J. R., OPPENHEIMER, M. and STEINER, J. E. (1979). *Chemical Senses and Flavour*, **4**, 49–61.
18. ROZIN, P. (1976). In: *Appetite and Food Intake*, Ed. Silverstone, T. Dahlem Konferenzen, Berlin, 285–312.
19. MOOK, D. G. (1963). *J. Comp. Physiol Psychol.*, **56**, 645–59.
20. EPSTEIN, A. and TEITELBAUM, P. (1962). *J. Comp. Physiol. Psychol.*, **55**, 753–9.
21. JACOBS, H. L. and SHARMA, K. N. (1969). *Ann. NY. Acad. Sci.*, **157**, 1084–125.
22. GARCIA, J., HANKINS, W. G. and RUSINIAK, K. W. (1974). *Science*, **185**, 823–31.
23. MAYER-GROSS, W. and WALKER, J. W. (1946). *Br. J. Exp. Pathol.*, **27**, 297–8.
24. NACHMAN, M. (1962). *J. Comp. Physiol. Psychol.*, **55**, 1124–9.
25. NACHMAN, M. and COLE, L. P. (1971). In: *Handbook of Sensory Physiology*, Vol. 4, Part 2. Ed. Beidler, L. M., Springer-Verlag, New York, 337–62.
26. SCOTT, E. M. and QUINT, E. (1946). *J. Nutr.*, **32**, 113–9.
27. NAIM, M., KARE, M. R. and INGLE, D. E. (1977). *J. Nutr.*, **107**, 1652–8.
28. RACKIS, J. J. (1974). *J. Am. Oil. Chem. Soc.*, **51**, 161A–74A.
29. BEAUCHAMP, G. K. and MALLER, O. (1977). In: *The Chemical Senses and Nutrition*. Eds Kare, M. R. and Maller, O. Academic Press, New York, 291–311.
30. MOSKOWITZ, H. R., KUMRAIAH, V., SHARMA, K. N., JACOBS, H. L. and SHARMA, S. D. (1975). *Science*, **190**, 1217–18.
31. MOSKOWITZ, H. R., KUMRAIAH, V., SHARMA, K. N., JACOBS, H. L. and SHARMA, S. D. (1976). *Physiol. Behav.*, **16**, 471–5.
32. LONDON, R. M., SNOWDON, C. T. and SMITHANA, J. M. (1979). *Physiol. Behav.*, **22**, 1149–55.
33. GALEF, B. G., JR. and CLARK, M. M. (1972). *J. Comp. Physiol. Psychol.*, **78**, 220–5.
34. NISBETT, R. (1969). *J. Per. Soc. Psychol.*, **10**, 107–16.
35. RODIN, J. (1976). In: *Hunger: Basic Mechanisms and Clinical Implications*, Eds Novin, D., Wyrwica, W. and Bray, G., Raven Press, New York, 409–19.
36. GRINKER, J., PRICE, J. M. and GREENWOOD, M. R. C. (1976). In: *Hunger: Basic Mechanisms and Clinical Implications*, Eds Novin, D., Wyrwica, W. and Bray, G., Raven Press, New York, 441–57.
37. TROSTLER, N. (1980). In: *Nutrition, Physiology and Obesity*, Ed. Schemmel, R., CRC Press, Inc. Cleveland, Ohio.
38. BRAY, G. A., YORK, D. A. and SWERLOFF, R. S. (1973). *Metabolism*, **22**, 435–42.
39. CHOCHINOV, R. H., ULLYOT, L. E. and MOORHOUSE, J. A. (1972). *N. Eng. J. Med.*, **286**, 1233–7.

40. LAWSON, W. B., ZEIDLER, A. and RUBENSTEIN, A. (1979). *Psychosom. Med.*, **41**, 219–27.
41. KAKOLEWSKI, J. W. and VALENSTEIN, E. S. (1969). *J. Comp. Physiol. Psychol.*, **68**, 31–7.
42. SCHELLING, J. H., TETREAUT, L., LASANGA, L. and DAVIS, M. (1965). *Lancet*, **1**, 508–12.
43. NIKI, A. and NIKI, H. (1975). *Lancet* **2**, 658.
44. NOVIN, D. (1976). In: *Hunger: Basic Mechanisms and Clinical Implications*, Eds Novin, D., Wyrwica, W. and Bray, G. A. Raven Press, New York, 357–67.
45. BOOTH, D. A. (1978). *Proc. Nutr. Soc.*, **37**, 181–91.
46. LEMAGNEN, J. (1978). In: *Recent Advances in Obesity Research*, Vol. 2, Ed. Bray, G., Newman, London, 45–53.
47. JORDAN, H. A., (1969). *J. Comp. Physiol. Psychol.*, **68**, 498–506.
48. JANOWITZ, H. D. and GROSSMAN, M. (1949). *Am J. Physiol.*, **159**, 143–8.
49. NAIM, M. and KARE, M. R. (1977). In: *The Chemical Senses and Nutrition*, Eds Kare, M. R. and Maller, O., Academic Press, New York, 145–62.
50. NAIM, M., KARE, M. R. and INGLE, D. E. (1978). *Life Sci.*, **23**, 2127–36.
51. KRATZ, C. M., LEVITSKY, D. A. and LUSTICK, S. (1978). *Physiol. Behav.*, **20**, 665–7.
52. GENTILE, R. L. (1970). *Physiol. Behav.*, **5**, 311–6.
53. KRATZ, C. M. and LEVITSKY, D. A. (1978). *Physiol. Behav.*, **21**, 851–4.
54. NAIM, M., BRAND, J. G., KARE, M. R., KAUFMANN. N. A. and KRATZ, C. M. (1980). *Physiol. Behav.*, **25**, 609–15.
55. KENNEY, J. J. and COLLIER, R. (1976). *J. Nutr.*, **106**, 388–91.
56. CAMPBELL, R. G., HASHIM, S. A. and VAN ITALLIE, T. B. (1971). *New Engl. J. Med.*, **285**, 1402–7.
57. WOOLEY, O. W. (1971). *Psychosom. Med.*, **33**, 436–44.
58. LEMAGNEN, J. (1967). In: *Handbook of Physiology. Alimentary Canal*. Sec. 6, Vol. 1, Ed. Code, C. F., American Physiological Society, Washington, DC, 11–30.
59. BOOTH, D. A. (1972). *J. Comp. Physiol. Psychol.*, **81**, 475–81.
60. SCLAFANI, A. and SPRINGER, D. (1976). *Physiol. Behav.*, **17**, 461–71.
61. HERBERG, L., DÖPPEN, W., MAJOR, E. and GRIES, F. A. (1974). *J. Lipid Res.*, **15**, 580–5.
62. SCLAFANI, A. (1978). In: *Recent Advances in Obesity Research*, Vol. 2, Ed. Bray, G., Newman, London, 123–32.
63. PAVLOV, I. P. (1910). *The Work of the Digestive Glands*, Translated by Thompson, W. T., Griffin, London.
64. GROSSMAN, M. I. (1967). In: *Handbook of Physiology*, Vol. 2 Ed. Code, C. F., American Physiological Society, Washington, DC, 835–63.
65. MAYER, G., ARNOLD, R., FEURLE, G., FUCHS, K., KETTERER, H., TRACK, N. S. and CREUTZFELDT, W. (1974). *Scan. J. Gastroenterol.*, **9**, 703–10.

66. PRESHAW, R. M., COOK, A. R. and GROSSMAN, M. I. (1966). *Gastroenterology*, **50**, 171–8.
67. STENQUIST, B., NILSSON, G., REHFELD, J. F. and OLBE, L. (1979). *Scan. J. Gastroenter.*, **14**, 305–11.
68. MOOR, J. G. and MOTOKI, D. (1979). *Gastroenterology*, **76**, 71–5.
69. CRITTENDEN, P. J. and IVY, A. C. (1937). *Am. J. Physiol.*, **119**, 724–33.
70. BEHRMAN, H. R. and KARE, M. R. (1968). *Proc. Soc. Exp. Biol. Med.*, **129**, 343–6.
71. NAIM, M., KARE, M. R. and MERRITT, A. M. (1978). *Physiol. Behav.*, **20**, 563–70.
72. THOMAS, J. E. (1941). *Proc. Soc. Exp. Biol. Med.*, **46**, 260–1.
73. RUDICK, J. and DREILING, D. A. (1968). *Surgery*, **63**, 683–5.
74. LANGLÓIS, K. J., LIM, R. K., ROSIER, G., STEWART, D. I. and STUMPFF, D. L. (1952). *Fed. Proc. Fed. Am. Soc. Exp. Biol.*, **11**, 88–9.
75. HARPER, A. A., KIDD, C. and SCRATCHERD, T. (1959). *J. Physiol. (London)*, **148**, 417–36.
76. SCHWARTZ, T. W., STENQUIST, B. and OLBE, L. (1979). *Scan. J. Gastroenter.*, **14**, 313–20.
77. OHARA, I., OTSUKA, S. and YUGARI, Y. (1979). *J. Nutr.*, **109**, 2098–105.
78. SMITH, C. A., HOLDEN, R. C. and HAWK, P. B., (1920). *Proc. Soc. Exp. Biol. Med.*, **17**, 98.
79. LEVEILLE, G. A., (1970). *Fed. Proc.*, **29**, 1294–301.
80. NAIM, M., KARE, M. R. and BRAND, J. G. (1980). In: *Olfaction and Taste*, Vol. VII, Ed. van der Starre, H. IRL Press Ltd., London & Washington DC, 426.
81. WOLF, S. G. and WOLFF, H. G. (1943). *Capitalized Human Gastric Function: An Experimental Study of a Man and His Stomach*, Oxford University Press, London.
82. HOLLANDER, F., SOBER, H. A. and BANDES, J. (1955). *Ann. N.Y. Acad. Sci.*, **63**, 107–19.
83. DUPRÉ, J. (1970). *Annu. Rev. Med.*, **21**, 299–316.
84. MCINTYRE, N., HODSWORTH, C. D. and TURNER, D. S. (1964). *Lancet*, **2**, 20–21.
85. PERLEY, M. J. and KIPNIS, D. M. (1967). *J. Clin. Invest.*, **46**, 1954–62.
86. FISCHER, U., HOMMEL, H., ZIEGLER, M. and JUTZI, E. (1972). *Diabetologia*, **8**, 385–90.
87. STEFFENS, A. B. (1976). *Am. J. Physiol.*, **230**, 1411–5.
88. NICOLAIDIS, S. (1969). *Ann. N.Y. Acad. Sci.*, **157**, 1176–1203.
89. NICOLAIDIS, S. (1977). In: *The Chemical Senses and Nutrition*, Eds Kare, M. R. and Maller, O., Academic Press, New York, 123–40.
90. STRUBBE, J. H. and STEFFENS, A. B. (1975). *Am. J. Physiol.*, **229**, 1019–22.
91. DEJONG, A., STRUBBE, J. H. and STEFFENS, A. B. (1977). *Am. J. Physiol.*, **233**, E380–8.
92. HELLEKANT, G. (1980). In: *Carbohydrate Sweeteners in Foods and Nutrition*, Eds Koivistoinen, P. and Hyvönen, L., Academic Press, London, 77–85.

93. BERRIDGE, K., GRILL, H. J. and NORGREN, R. (1981). *J. Comp. Physiol. Psychol.*, **95**(3), 363–82.
94. WOODS, S. C., HUTTON, R. A. and MAKOUS, W. (1970). *Proc. Soc. Exp. Biol. Med.*, **133**, 964–8.
95. SIEGEL, S. (1975). *J. Comp. Physiol. Psychol.*, **89**, 189–99.
96. GOLDFINE, I. D., ABRAIRA, C., GRUENWALD, D. and GOLDSTEIN, M. S. (1970). *Proc. Soc. Exp. Biol. Med.*, **133**, 274–6.
97. SJÖSTRÖM, L., GARELLICK G., KROTKIEWSKI, M., and LUYCKX, A. (1980). *Metabolism*, **29**, 901–9.
98. LOUIS-SYLVESTRE, J. (1976). *Am. J. Physiol.*, **230**, 56–60.
99. DEUTSCH, R. (1974). *J. Comp. Physiol. Psychol.*, **86** 350–8.
100. VALENSTEIN, E. S. and WEBER, M. L. (1965). *J. Comp. Physiol. Psychol.*, **60**, 443–6.
101. HOEBEL, G. B. and TEITELBAUM, P. (1966). *J. Comp. Physiol. Psychol.*, **61**, 189–93.

10

Chemical and Technological Modification of Physiological Effects in Food Carbohydrates

M. W. KEARSLEY and R. H. P. LIAN-LOH
*National College of Food Technology,
University of Reading,
Weybridge, Surrey, UK*

ABSTRACT

Sugars are widely used in foods especially for their sweetening ability, although in recent years a number of bodily disorders have been associated with these versatile food components. Such disorders include tooth decay, the most prevalent disease in the Western world, obesity, heart disease and diabetes mellitus. Whilst sucrose has largely received the blame for these disorders it is not the only carbohydrate which may be implicated since others are commonly used in or are naturally present in foods. Generally however, replacement of sucrose by apparently safer sugars is often seen as advantageous.

Carbohydrates possess other useful physico-chemical and physiological properties in addition to the sweet function and it is likely that in the future modification of carbohydrates to control these properties will become very important. Of the recent developments in this area, the most promising to the food manufacturer is the production of chemically modified glucose syrups, in particular by hydrogenation. Such syrups are already known to possess certain advantages with regard to fermentability, browning ability and metabolic response, and current research may reveal other useful properties. Although hydrogenated glucose syrups are not yet permitted as food additives in the UK it is envisaged that future legislation will enable their novel physico-chemical and physiological properties to be utilised in foods for the benefit of the consumer.

INTRODUCTION

Sugar, or more specifically sucrose, is used in foods primarily for its sweetening ability. However in recent years it has been associated with a number of bodily disorders.[1-3] These include tooth decay, the most prevalent disease in the Western world, obesity, heart disease and diabetes mellitus. The evidence that sucrose alone is responsible for these conditions is by no means conclusive in all respects and refined carbohydrate in general does not stand alone as being potentially dangerous to the human body. Indeed the seemingly harmless potato has been shown to be harmful to rats fed the raw starch, the resulting large caecum eventually causing asphyxiation in the animal.[4]

Sucrose is not the only sugar in common use in foods and whilst many foods contain sucrose, it may not be the sugar present in the highest concentration. Furthermore, some of the harmful effects of sucrose are also shown by other carbohydrates.

Very few foods do not contain carbohydrate in one form or other and because of the versatile nature of the carbohydrates we could scarcely exclude them from foodstuffs without losing palatability. Useful properties of sugars include: sweetness, which is probably one of their most important assets; their preservative effect, as in jams; and their browning abilities. Carbohydrate may also be used in the treatment of certain metabolic disorders and as an easily assimilable source of energy in illness.

Fructose is generally considered to be the sweetest common sugar with sucrose close behind and with the latest synthetic sweeteners some 3000 times sweeter than sucrose.[5] Glucose syrups, excluding high fructose syrups, are by comparison, even at maximum conversion, only about half as sweet as sucrose.[6,7] Carbohydrates derived from starch, as opposed to sucrose or synthetic compounds, have more aesthetic appeal and glucose syrups possess certain metabolic advantages compared with sucrose as well as being cheaper to produce.[8-13] These versatile compounds have been used for many years in the food industry and the recently developed high fructose syrups are in great demand for use in soft drinks.

For food use the physico-chemical properties of a carbohydrate are probably of prime importance although the physiological effects should obviously not be neglected. It is the object of this paper to describe some of the metabolic effects of ingested carbohydrate.

EFFECTS OF COMMON CARBOHYDRATES ON BLOOD GLUCOSE ELEVATION

A simple experiment was carried out[14] whereby human volunteers, after an overnight fast and fasting blood sample, ingested a single 50 g dose of either invert sugar, sucrose, fructose or isomerised syrup. Glucose was taken for comparison purposes. No significant differences were established between the blood glucose profiles of invert sugar, sucrose, isomerised syrup and glucose, whilst fructose gave a significantly lower average peak response. Individual blood glucose responses to fructose revealed a wide variation between subjects, some giving elevated values, othern no response at all, which was in agreement with previous work.[15,16]

When 50 g of the high fructose syrups (containing about 25 g glucose and 25 g fructose) were subsequently ingested by other subjects, in every case the fructose syrups produced higher peak blood glucose values then 50 g glucose alone. This indicated that probably within certain limits, peak values for blood glucose were independent of the dose of ingested glucose (later work by other workers[17] confirmed this). To clarify these results, further glucose tolerance tests were carried out using reduced loads of glucose and high fructose syrups.[18] The results indicated that there was no significant difference in blood glucose elevation between 25 g and 50 g glucose and 20 g high fructose syrup (containing about 10 g glucose). Since 10 g glucose gave a significantly lower peak glucose value than these three test carbohydrates, the results indicate some conversion of fructose to glucose or suppression of insulin by the fructose when high fructose syrups are ingested. Further investigation is obviously required to clarify this situation.

When commercially produced glucose syrups are ingested there is no significant difference in blood glucose profile or serum insulin profile between a syrup of any DE and the glucose standard suggesting that stomach emptying and hydrolysis are not limiting factors and that absorption controls blood glucose peak height after ingestion.

PHYSICO-CHEMICAL MODIFICATION OF CARBOHYDRATES

Physico-chemical modifications to carbohydrates may be carried out to modify the physiological response to the carbohydrate; some modifications will have a more severe effect than others. For the purposes of this paper physical modification has been limited to the fractionation of glucose syrups by reverse osmosis.[19, 20] Theoretically, using this technique, glucose syrups can be produced with a limited carbohydrate spectrum, which might influence the physiological response after ingestion. Unfortunately, although physico-chemical properties of the syrups can be changed slightly by reverse osmosis fractionation, no differences were shown in their physiological properties as judged by glucose tolerance tests, although long term trials indicated some differences.[8, 21] It is envisaged, however, that when membrane technology is further advanced, glucose syrups may be completely fractionated into their component parts and the rate of hydrolysis of these may have some effect on the rate of entry of glucose into the bloodstream.

It is in the area of chemical modification of carbohydrates that physiological response may most easily be changed. Many types of modification are possible and the carbohydrate may be partly or wholly derivatised. Acetylation and benzoylation of lower molecular weight carbohydrates result in bitter derivatives of varying water solubility whilst acetylated and benzoylated glucose syrups are virtually water-insoluble and hence of little use in foods. The main use of this type of derivative has been found to be in the preparation of pure glucose oligomers for GLC or HPLC.[22]

Oxidation is likely to offer some possibilities although the process is not at present easy to carry out and toxicity problems may arise.[23] Hydrogenation seems to be the technique most likely to be acceptable to the consumer since in effect it is only changing one natural product for another, that is sorbitol for glucose. Sorbitol can be produced from D-glucose by the addition of one molecule of hydrogen using a temperature of 100°C, a hydrogen pressure of 100 atm and a nickel catalyst. Sorbitol and glucose have several differences in their properties and in the context of this paper the main difference is that sorbitol has a low insulin demand, hence its use in diabetic foods; unfortunately it also has a powerful laxative effect in large doses.

By treating glucose syrups in the same way as glucose, a similar reaction occurs. The glucose component is converted to sorbitol, maltose to maltitol, maltotriose to maltotriitol and so on. It is of course only the terminal glucose residue in each oligomer which is affected by hydrogenation. Thus it is obvious that the greatest changes in properties after hydrogenation will be observed in the highest DE syrups. Although the physiological implications have not as yet been considered, partially hydrogenated syrups provide some benefits to the consumer from a physico-chemical aspect.

Hydrogenation causes no change in viscosity, osmotic pressure or threshold sweetness concentration (although at supra-threshold concentrations hydrogenated syrups are sweeter[7]). Specific rotation, Maillard browning, fermentability and blood glucose elevation decrease.[24] This latter property is of course the most important consideration at the present time.

SHORT TERM STUDIES

A study was carried out in which subjects ingested 50 g of one of seven test carbohydrates after an overnight fast. These were glucose, maltitol, glucose/maltitol (50:50) and hydrogenated 21 DE, 43 DE, 65 DE and high maltose glucose syrups. Capillary blood samples were taken prior to ingestion and for up to 2 h after ingestion and blood glucose values determined by the GOD-Perid method. Several observations were made from this study. Maltitol and hydrogenated 65 DE glucose syrup gave significantly lower blood glucose values 30 min after ingestion than the other samples and there was no significant difference between the two. All subjects complained of diarrhoea after ingesting the hydrogenated 65 DE syrup but this was not unexpected owing to its high sorbitol content. Two subjects had similar trouble with the hydrogenated 43 DE syrup but no problems were reported with the remaining samples.

When these results are compared with those for different loadings of glucose, the blood glucose peak value obtained for maltitol is unusually low. On hydrolysis 50 g maltitol yields approximately equal quantities of glucose (25 g) and sorbitol (25 g) and previously 25 g glucose was shown to give the same blood glucose peak as 50 g glucose. However, 50 g maltitol gives a significantly lower peak value than 50 g glucose. None of the subjects complained of diarrhoea with maltitol, implying that it is utilised and does not cause osmotic diarrhoea. Possible mechanisms to explain the reduced peak height include the following.

1. The maltitol is partially hydrolysed and partly absorbed (as an intact molecule) into the bloodstream from the gut.
2. The maltitol is slowly hydrolysed and thus peak values for blood glucose are only slowly attained.
3. The maltitol is hydrolysed as normal maltose would be and glucose transport is inhibited by sorbitol.
4. The maltitol is hydrolysed only partly and the main bulk is utilised in the lower gut by the microflora.

LONG TERM STUDIES

To clarify the situation the work was taken a stage further and a prolonged feeding trial using maltitol was carried out. In all, four carbohydrates were evaluated: glucose as the standard, 43 DE glucose syrup, maltitol, and Lycasin®, a commercially hydrogenated glucose syrup similar to a high maltose syrup.

Each trial lasted 15 days. Five subjects were involved and an initial glucose tolerance test was carried out on day 1 using one of the test carbohydrates. After an overnight fast a 10 ml fasting blood sample was taken and the subject presented with a solution of the carbohydrate at a concentration of 0·5 g/kg body weight dissolved in 4 ml of water/kg body weight. After ingestion, blood samples were taken every 30 min for a total period of 3 h. On days 1–4 a normal diet was followed and the trial proper started on day 5. On days 5–14 inclusive, each subject ingested approximately 2000 cal/day, 40% of the calories arising from the carbohydrate under test and 60% from protein. Fat and other carbohydrate were wherever possible excluded from the diet. On days 8 and 14 further tolerance tests were carried out. All blood samples were immediately centrifuged and the separated serum frozen. Subjects on the trial consuming maltitol and Lycasin were also required to collect all urine and faeces samples.

At the end of the trial, all blood samples were analysed for glucose, insulin, sorbitol, maltitol, triglycerides, cholesterol and high density cholesterol. All faeces and urine samples were analysed as soon after collection as possible for sorbitol and maltitol. After a suitable rest, further trials were carried out until each subject had ingested each of the test compounds.

No discomfort was experienced by any of the subjects when

consuming either glucose or 43 DE glucose syrup. All subjects complained of some stomach pains, diarrhoea and flatulence initially when consuming maltitol and Lycasin but after 3-4 days the effects were either much reduced or disappeared completely, presumably because of the adaptation to the carbohydrate. This in fact would be expected from any unaccustomed carbohydrate in the diet, which was included suddenly at high concentrations

Blood glucose profiles Maltitol and Lycasin gave significantly lower blood glucose peak values than did glucose and 43 DE glucose syrup. During the course of the trial a gradual adaption to maltitol and Lycasin occurred with peak values being higher after 14 days than at the start of the experiment. There was no significant difference between glucose and 43 DE glucose syrup at any time or between maltitol and Lycasin at any time.

Serum insulin profiles Maltitol and Lycasin gave significantly lower serum insulin peak values than did glucose and 43 DE glucose syrup. The gradual adaptaticn to Lycasin and maltitol was shown by a slight increase in peak values after 14 days of the trial. There was no significant difference between glucose and 43 DE glucose syrup at any time and Lycasin gave a significantly higher peak value than did maltitol at each test during the course of the trial. This would be predictable from the fact that Lycasin, on breakdown, yields more glucose than does maltitol.

Sorbitol and maltitol No sorbitol or maltitol was detected in the blood of any of the subjects on any of the diets.

Serum triglycerides All carbohydrates induced a fall in serum triglycerides and no significant difference was established between samples.

Serum cholesterol and high density cholesterol None of the samples induced any change in these parameters.

Analysis of urine No sorbitol or maltitol was detected in the urine of subjects on the glucose and 43 DE glucose syrup diets. Subjects on both Lycasin and maltitol diets excreted on average 10 g sorbitol and between 0·25 g and 0·5 g maltitol over the 10-day trial; there was no significant difference between Lycasin and maltitol.

Analysis of faeces No sorbitol or maltitol was detected in the faeces of the subjects on the glucose or 43 DE glucose syrup diets. There was no significant difference between Lycasin and maltitol, the subjects excreting between 0·1 and 8 g of sorbitol and up to 11 g of maltitol over the 10-day trial whether on the Lycasin or maltitol diets.

Total carbohydrate intake in the 10 days was about 2 kg and percentage retention by each subject was over 99%. The analysis of urine and faeces shows: (i) that very little Lycasin or maltitol is lost from the body, and (ii) that maltitol can be absorbed as an intact molecule into the body owing to its presence in the urine.

CONCLUSIONS

With the exceptions of sorbitol and fructose, most common carbohydrates induce similar blood glucose profiles after ingestion. By chemical modification or more specifically, by hydrogenation, this response can be modified and compounds produced with lowered insulin demands in addition to the many useful physico-chemical properties these modified sugars possess. Such modified sugars could be used with advantage in diabetic foods.

ACKNOWLEDGEMENTS

The authors wish to thank Roquette Frères for financial assistance to carry out the hydrogenation studies, and also the many subjects who took part in the physiological studies.

REFERENCES

1. VISSER, W. (1979). In: *Sugar: Science and Technology*, Eds Birch, G. G. and Parker, K. J. Applied Science Publishers, London, 457–64.
2. YUDKIN, J. (1979). In: *Sugar: Science and Technology*, Eds Birch, G. G. and Parker, K. J. Applied Science Publishers, London, 425–35.
3. MACDONALD, I. (1979). In: *Sugar: Science and Technology*, Eds Birch, G. G. and Parker, K. J. Applied Science Publishers, London, 415–23.
4. EL-HARITH, E. A., WALKER, R., BIRCH, G. G. and SUKAN, G. (1977). *Fd Chem.*, **2**(4), 279.
5. PANCOAST, H. M. and JUNK, W. R. (1980). *Handbook of Sugars*, 2nd edn, Avi Publishing Co., Westport, Conn.
6. KEARSLEY, M. W., BIRCH, G. G. and DZIEDZIC, S. Z. (1978). *Lebensm. -Wiss. u Technol.*, **11**, 23.

7. KEARSLEY, M. W., DZIEDZIC, S. Z., BIRCH, G. G. and SMITH, P. D. (1980). *Starke*, **32**(7), 244.
8. BIRCH, G. G., ETHERIDGE, I. J. and GREEN, L. F. (1973). *Br. J. Nutr.*, **29**, 87.
9. AHRENS, E. H., HIRSCH, J., OETTLE, K., FARQUAR, J. W. and STEIN, Y. (1961). *Trans. Ass. Am. Phycns.*, **74**, 134.
10. MACDONALD, I. (1974). *The Practitioner*, **212**, 448.
11. NAISMITH, D. J. and RANA, I. A. (1974). *Nutr. Metabol.*, **16**, 285.
12. AHRENS, R. A. (1974). *Am. J. Clin. Nutr.*, **27**, 403.
13. KEARSLEY, M. W., BIRCH, G. G., TABIRI, J. N. and DUDBRIDGE, M. J. (1980). *Starke*, **32**(6), 205.
14. BIRCH, G. G. and KEARSLEY, M. W. (1977). *Starke*, **29**(10), 348.
15. CORNBLATH, M., ROSENTHAL, I. M., REISNER, S. H., WYBREGT, S. H. and CRANE, R. K. (1963). *New Engl. J. Med.*, **269**, 1271.
16. MATSCHINSKY, F. M., LANDGRAF, R., ELLERMAN, J. and KOTLERBRAJTBURG, J. (1972). *Diabetes*, **21** (Suppl. 2), 555.
17. MACDONALD, I., KEYSER, A. and PACY, E. (1978). *Am. J. Clin. Nutr.*, **31**(8), 1305.
18. KEARSLEY, M. W. and BIRCH, G. G. (1976). *Proc. Nucr. Soc.*, **36**, 45A.
19. KEARSLEY, M. W. (1976). *Starke*, **28**, 138.
20. BIRCH, G. G. and KEARSLEY, M. W. (1974). *Starke*, **26**, 220.
21. KEARSLEY, M. W., FAIRHURST, E. and GREEN, L. F. (1975). *Proc. Nutr. Soc.*, **34**, 60A.
22. DZIEDZIC, S. Z. and KEARSLEY, M. W. (1978). *J. Chromatog.*, **154**, 295.
23. BOOK, L. S., HERBERT, J. J. and STEWART, D. (1978). *J. Pediat.*, **92**(5), 793.
24. KEARSLEY, M. W. and BIRCH, G. G. (1979). In: *Sugar: Science and Technology*, Eds Birch, G. G. and Parker, K. J. Applied Science Publishers, London, 287–309.

11

Sugars and Dental Caries

W. M. EDGAR

University of Newcastle upon Tyne,
Dental School,
Newcastle upon Tyne, UK

ABSTRACT

Dental caries is believed to result from dissolution of the mineral part of the tooth during the 15–20 min period after taking sugar-containing foods, confectionery and beverages, when the pH is low owing to metabolism of the sugar to acids by the bacterial flora of the tooth surface (dental plaque).

Evidence concerning the relative caries-producing potential (cariogenicity) of sugars, starch and various nutritive sweeteners derives from human epidemiology and clinical trials, animal caries tests and in vivo and in vitro biochemical tests. While clinical testing in man is the only completely valid method of assessing cariogenicity, it is difficult and costly. One convenient alternative screening method involves determination of pH in dental plaque after volunteers consume dietary substrates or products.

Present information suggests that sucrose, while not necessarily more cariogenic than glucose, is the main dietary cause of caries in developed societies owing to its ubiquity. The cariogenicity of fructose is somewhat less than that of sucrose, but greater than that of lactose and maltose; however, the cariogenicity of a mixture of glucose and fructose may equal that of sucrose. Corn syrups are likely to vary in cariogenicity in proportion to their mono- and disaccharide content; starch itself gives rise to little if any caries. Sorbitol, mannitol, sorbose and chlorosugars do not support rapid acid production by the dental flora and have a low cariogenicity in other test systems; total replacement of sucrose in the human diet by xylitol abolished new caries over 2 years. Cariogenicity of hydrogenated starch products is probably low but may vary between sources.

Besides possible substitution of sucrose by less-cariogenic sweeteners other approaches such as altered patterns of food use, and protective additives, may assist in developing ways to reduce dental disease resulting from misuse of sugar.

INTRODUCTION

Dental caries, a disease which afflicts virtually all the population of developed societies, is geographically and historically associated with sugar consumption.[1] The reason for this association is almost universally believed to lie in the nature of the caries process itself, involving as it does the dissolution of the mineral phase (hydroxyapatite) of the dental enamel and underlying dentine in conditions of low pH. These conditions are the result of the formation of acids, mainly lactic, during the anaerobic fermentation of carbohydrates by the bacteria inhabiting the teeth—the dental plaque. When the pH of the environment of the tooth falls, the degree of saturation with respect to various calcium phosphates is altered in such a way as to increase the driving force towards demineralisation; at some pH there will presumably be a break point below which demineralisation begins, and this hypothetical pH is termed the critical pH. It follows therefore that the number of times in a day that the pH of the plaque falls below the critical value, and the duration and extent of such falls in pH, will determine the rate of demineralisation and so of caries formation. This 'acid theory' of caries receives overwhelming support from numerous studies of various types made over many decades, although the possibility of an additional slow demineralisation of the tooth by calcium complexes which may form at neutrality cannot be wholly disregarded.

The rapidity of the pH changes which may occur in plaque are shown in Fig. 1;[2] this rapidity suggests that the caries process occurs briefly after each intake of readily fermentable carbohydrate and is interspersed with periods during which a limited remineralisation by calcium and phosphate species in the environment of the tooth may occur. The significance of sugars in caries formation lies mainly in the ease with which they are fermented by the plaque microorganisms. Table 1 shows data for the rate of acid production from common dietary carbohydrates by plaque homogenates *in vitro*.

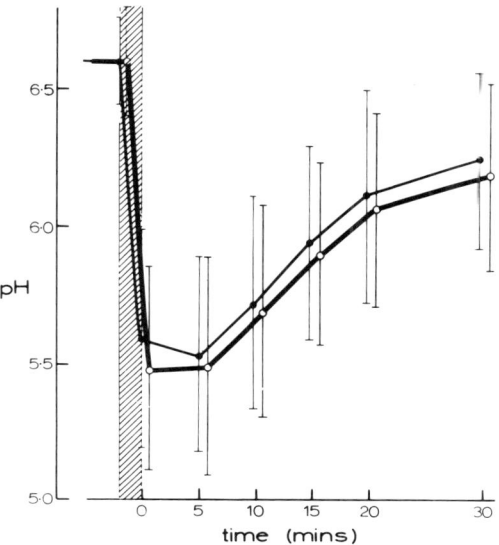

FIG. 1. Typical changes in plaque pH following mouth-rinsing with 10% solutions of glucose (thin line) or sucrose (thick line). Saliva pH: glucose, 6·69 ± 0·46; sucrose, 6·61 ± 0·45. Saliva flow: glucose, 0·90 ± 0·43; sucrose, 0·99 ± 0·40. Two minute mouth-rinsing denoted by hatched area. Mean of 33 experiments in all 11 subjects. Data from Reference 2.

TABLE 1

ACID PRODUCTION BY SUSPENSIONS OF PLAQUE INCUBATED WITH VARIOUS SUGARS AND SUGAR SUBSTITUTES AT pH 6·8 (DATA FROM REFERENCE 3)

Substrate	Acid production rate[a]
Glucose	100
Invert sugar, sucrose	100
Glucose syrups (DE 40, 60)	100
Fructose	80–100
Lycasin (low maltitol)	50–70
Lactose	40–60
Lycasin (high maltitol)	20–40
Maltitol, sorbitol	10–30
Mannitol, xylitol	0

[a] nmols NaOH required for neutralisation/mg plaque/min, expressed as percentage of glucose control.

In addition to serving as a source of acid, however, sugars may be utilised in the synthesis by the bacteria of extracellular polysaccharides, which are believed to be important in allowing the bacteria to attach to and cohere on the tooth surface. Extracellular glucans formed specifically from sucrose by many plaque bacteria, especially streptococci, have been extensively studied; the fact that sucrose is the substrate for the enzymes (glucosyl transferases) involved in glucan synthesis has led to the concept that this sugar may be specifically involved in the aetiology of caries.

At this point it is appropriate to consider how evidence concerning the potential for producing caries (cariogenicity) of dietary items or constituents is gathered. Completely valid evidence is of course obtained only by counting cavities, either measuring caries in a population and relating this to dietary practices or following the effect on caries incidence of dietary changes introduced experimentally. Since caries is slow in progression from the initial lesion to the clinically detectable cavity, such surveys are time-consuming and expensive to carry out, and can only be relied on for information about major dietary differences, not minor details.

Cariogenicity may also be tested in experimental animals, usually rats, to which foods or substrates are fed at high concentration and frequency over a period of weeks after which their teeth may be sectioned and cavities counted. In order to produce caries rapidly the animals' diets usually contain high levels of sugar—up to 56%— and comparisons between different foods or sweeteners may be difficult to interpret in view of the extreme nutritional conditions. The most up to date animal caries tests avoid some of these difficulties by employing a programmed feeding machine to give a fixed amount of test foodstuff at a pre-set frequency; meanwhile the animals receive their basic nutrients by stomach tube.[4, 5] Normal nutrition is thus assured while the only foodstuff to contact the teeth is that under test.

Animal tests cannot wholly replace human experiments, however, since factors such as the method of food use may be very important in determining cariogenicity and are difficult if not impossible to reproduce in animals. Human 'guinea-pigs' can provide valuable information on caries-related properties of food, especially their potential fermentability by the plaque during normal use. Measurement of plaque pH before and after eating or drinking, or mouth-rinsing with a sugar solution provides an important source of information on the cariogenic potential of foods,[6-9] although other factors (e.g. the effect of the food on tooth solubility) also need to be considered. The softening (demineralisation) of a slab of enamel worn in the mouth and harbouring a bacterial plaque, when exposed frequently to sugars or sweeteners (Table 2) provides a measurement which might be expected to represent the disease process more closely.[10]

TABLE 2
ESTIMATED CONTRIBUTION OF NINE TEST SUBSTRATES TO SOFTENING (DEMINERALISATION) OF BOVINE ENAMEL SLABS COMPARED WITH SUCROSE CONTROL (DATA FROM REFERENCE 10)

Substrate	Percentage of control
Distilled water	8
Fructose	110
Glucose	117
Lactose	35
Mannitol	42
Melibiose	56
Raffinose	108
Sorbitol	45
Xylitol	−11
Xylose	−21

Softening measured as change in Knoop indenter penetration (μm) after wearing plaque-covered slabs mounted in a partial denture for 1 week. Slabs exposed to test substrate four times a day.

Additional information concerning cariogenicity may come from *in vitro* tests of acid production enamel solubility in the presence of substrates incubated with oral bacteria in pure culture or homogenates of plaque or saliva; the drawback of such tests, that they involve a closed system while in the mouth the system is open and therefore influenced by factors such as rates of diffusion of substrate to and acids from the plaque, has been partially overcome by ingenious devices to represent the conditions more closely—so-called 'artificial mouths'. However, the advantages of these devices may be outweighed by their complexity and simpler (and perhaps more valid) human plaque pH methods may be preferable.

CONVENTIONAL DIETARY SUBSTRATES

Having outlined the background to food cariogenicity and some of the methods of study available, the evidence regarding the contribution of common dietary carbohydrates to the incidence of caries can be discussed.

The Cariogenicity of Sucrose

Sucrose is readily metabolised to acid by the oral flora, and there is evidence that many oral species possess a specific uptake mechanism for sucrose and so prior hydrolysis of the sucrose by extracellular invertase is not required. The association between sucrose consumption and caries noted above stongly suggests that this sugar is responsible for the majority of caries found in developed societies (Fig. 2).

Numerous surveys of the prevalence of caries in groups of people

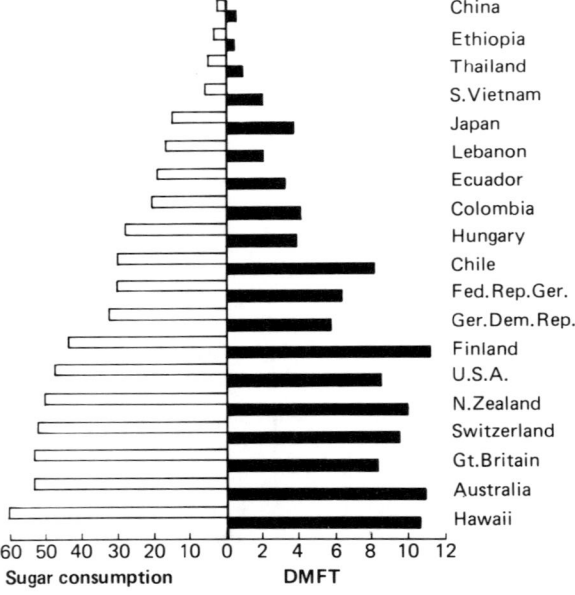

FIG. 2. Worldwide association between caries (Decayed, Missing and Filled Teeth) and sugar consumption. Data from Reference 1 (artwork kindly supplied by Dr A. J. Rugg-Gunn).

consuming varying levels of sugar have indicated a rise in caries associated with sugar consumption, although other dietary changes inevitably accompany the change in sugar level. For example, children living in a residential school in Australia, where dietary restriction of sugar was imposed, had very low caries experience compared with children in normal schools.[11] Sufferers of the rare disease of hereditary fructose intolerance, who must avoid dietary sources of fructose including sucrose, experience little caries.[12] The population of Tristan da Cunha had little dental decay until contact with the outside world led to the introduction into their diet of refined processed foods containing sucrose.[13] Consumption of sucrose-containing caramels by institutionalised subjects in the Vipeholm hospital in Lund, Sweden led to the development of caries (Table 3) when they were administered between meals.[14] Near total replacement of dietary sucrose by xylitol for 2 years (Table 4) greatly reduced the development of caries in a group of Finnish volunteers (the Turku study[15]).

As mentioned previously, the synthesis of extracellular glucans from sucrose is important for the attachment of oral bacteria to surfaces (especially *Streptococcus mutans*, a species which is asso-

TABLE 3

CARIES DEVELOPMENT (TOOTH SURFACES DECAYED PER PERSON PER YEAR) IN THE VIPEHOLM EXPERIMENT (DATA FROM REFERENCE 14, TABLE 12)

Group	Caries Development	
	A	B
Control (basal diet only)	0·3	
Sucrose (75 and 300 g/day, liquid)	0·8	
Bread, males (345 g high-sugar bread)	1·0	
Bread, females (345 g high-sugar bread)	0·7	
Chocolate (300 g sucrose with meals or 65 g chocolate between meals)	0·3	1·4
Caramel (345 g high-sugar bread with meals or 22 caramels between meals)	0·4	2·9
8-Toffee (+ sucrose solution with meals)	—	3·5
24-Toffee, males (+ sucrose solution with meals)	—	4·4
24-Toffee, females (+ sucrose solution with meals)	—	7·2

A, sugar supplement consumed with meals; B, sugar consumed with and between meals.

TABLE 4
CUMULATIVE DEVELOPMENT OF CARIES OVER 2 YEARS IN HUMAN SUBJECTS CONSUMING DIETS CONTAINING SUCROSE (S GROUP), XYLITOL (X GROUP) AND FRUCTOSE (F GROUP) (DATA FROM REFERENCE 15)

Group	C1+C2	Caries scores[a] + secondary	C2 only
S	7·2	10·5	3·3
X	0·0	0·9	1·5
F	3·8	6·1	3·6

[a] Caries increment scored as 'decayed, missing or filled surfaces' (DMFS). C1, precavitation stage (white spot); C2, cavitation stage; secondary, caries occurring round existing fillings. C2 caries in X group balanced by remineralisation of C1 lesions.

ciated with caries and which may play an important part in the cause of the disease because of its ability to form acid and grow at low pH). This suggested that sucrose might have a specific role in caries by contributing to the formation of a cariogenic plaque, and prompted many studies to compare the cariogenicity of sucrose with that of monosaccharides, usually glucose, which are not substrates for glucan synthesis. The results of experiments in rats are somewhat variable; overall they tend to suggest that the cariogenicities of sucrose and glucose are not dramatically different[16-18] although the trend is for sucrose to be more cariogenic.[19, 20] Similar levels of caries developed in monkeys fed diets rich in sucrose or invert sugar.[21] Children consuming between-meal foods containing invert sugar rather than sucrose developed 35% fewer new cavities than control subjects, but these children also restricted their total sugar intake and the effect of the sugar substitution alone cannot be assessed.[22]

Acid production *in vitro*, and plaque pH falls *in vivo* are similar (Fig. 1) whether glucose or sucrose is used as the substrate;[2, 3] comparable degrees of softening were observed in the surfaces of enamel slabs worn in the mouth of volunteers and periodically dipped in solutions of sucrose or glucose as shown in Table 2.[10] Demineralisation of enamel slices incubated with saliva and sucrose or glucose in an artificial mouth device known as the 'Orofax'[23] was not significantly different (Table 5).

Present evidence thus indicates that sucrose is unlikely to be more cariogenic in man than glucose. However, since sucrose is present at higher levels in the diet than any other simple sugar, it is presumed to be responsible for a large part of the decay occurring in developed societies.

TABLE 5
EFFECT OF CARBOHYDRATES ON ENAMEL DEMINERALISATION IN THE 'OROFAX' (DATA FROM REFERENCE 23)

Carbohydrate	Depth of demineralisation (μm)
Sucrose 20%	165·2
Glucose 20%	155·5
Lactose 20%	118·5^a
Starch 20%	84·2^a
Lactose 5%	44·1^b
Corn syrup 5%	44·8^b

Difference between values ^{a} and ^{b} is significant ($p < 0.01$).

The *amount* of sugar consumed is not, however, believed to be the controlling factor, but rather the *frequency* of consumption. This is indicated by the results of the Vipeholm study (Table 3) where subjects receiving between-meal sugar and experiencing a high level of caries received no more total sugar than controls who consumed their sugar at meal times.[14] The effect of frequency has been demonstrated in rat experiments using a programmed feeding machine delivering a variable number of meals per day—similar quantities of sugar give different levels of caries depending on the number of different occasions at which they are consumed.[4]

The frequency effect can be readily understood when the pH behaviour of plaque is considered. Each carbohydrate intake will be followed by a fall in pH and a demineralisation of the tooth—the more frequently this happens, the greater the caries. Plaque pH usually remains high while food is in the mouth because of the flow of alkaline saliva, so if all sugar is consumed at one meal time, then only one pH fall will occur. If a sugary food is consumed between two non-sugary foods in sequence, the fall in pH it normally would provoke is diminished.[24]

Some experiments with rats, using a homogenised diet consisting of various concentrations of sucrose, have suggested that a plateau in cariogenicity is reached at about 10–15% sucrose[25, 26] while other experiments indicate that caries continues to increase with rising sugar concentration up to 50–60%.[27, 28] These differences are probably attributable to variables in experimental design, and may not relate to human diets, since numerous factors contribute to a food's cariogenicity besides sugar concentrations. These factors may include consistency, stickiness (retention in the mouth), method of consumption (chewing, sucking, sipping, etc.), stimulus to salivation, and pH; several of these properties may interact in a complex fashion in determining the ability of a food to provoke a plaque pH fall.[7]

There is as yet no agreement on the minimum concentration of sucrose required to produce a fall in plaque pH to harmful levels. Results vary with the method used: a marked fall in pH was observed when subjects wearing a denture, carrying a built-in miniature electrode, rinsed their mouths with sucrose concentrations as low as 0·025%,[9] while similar levels did not provoke a fall in pH as measured in samples of plaque removed from the mouth at intervals after rinsing.[6]

Other Simple Sugars

In the study on monkeys cited earlier,[21] fructose was found to be slightly less cariogenic than either sucrose or invert sugar. This conclusion is supported by the Turku study, where for 2 years groups of volunteers consumed a diet in which sucrose was substituted with either xylitol or fructose (see below, Table 4). In rats, broadly comparable findings have been obtained.[16, 20]

Lactose and galactose have been shown to give a smaller drop in plaque pH than similar levels of sucrose, glucose or fructose.[9, 29] *In vivo* softening of enamel slabs was lower with lactose as substrate than with fructose or glucose,[10] and in an artificial mouth (Orofax) lactose gave significantly less demineralisation of enamel than comparable levels of sucrose.[23] Milk only provokes a small plaque pH drop, and is considered to have a low cariogenicity.[30, 31]

Starch and its Hydrolysis Products

Animal experiments have in general found that starch has a low cariogenicity[32–34] and this is borne out by the low caries experience of subjects consuming a diet low in sugar but with abundant starchy foods.[1, 11-15] However, plaque pH studies have indicated that a drop in pH can be observed with starch as substrate, indicating some acid production from the products of starch digestion by salivary α-amylase.[7, 9] Maltose (and probably maltotriose) can apparently be used for acid production by plaque[9] and maltose was found to give as much or more caries compared with sucrose or glucose when fed to rats.[18] On the other hand, starch hydrolysates may reduce bacterial colonisation of the teeth by inhibiting extracellular glucan synthesis.[35] Studies have shown that the low cariogenicity of wheat starch in animal experiments is not caused by the presence of an amylase inhibition from the endosperm.[36]

Corn syrups are likely to vary as to their cariogenicity with the degree of hydrolysis of the original starch to low molecular weight products (dextrose equivalents). Bacterial acid production *in vitro* from glucose syrups is high,[3] and animal experiments have shown that a glucose syrup containing approximately 33% glucose and maltose (DE 41) was about as cariogenic as sucrose when incorporated into the diet either as a spray-dried powder or in its original viscous liquid form.[37] Eating sweets containing spray-dried wheat starch hydolysate led to the accumulation of less plaque in human subjects than with sucrose-containing sweets[38] but the small difference would not be expected to have a clinically detectable effect. However, in the Orofax, a 5% solution of a corn syrup of unstated composition produced over 70% less enamel demineralisation than a 20% sucrose solution (Table 5). As consumption of corn syrups is increasing and their composition can be controlled, scope exists for examining in more detail their cariogenic potential.

NUTRITIVE SWEETENERS AND SUGAR SUBSTITUTION

The possibility of substituting sucrose with non-sugar sweeteners, especially in such foods as sweets, snacks and beverages used frequently between meals, with the aim of reducing the cariogenicity of such items has raised much interest and stimulated much research.

Xylitol

The most exhaustive work in this field was the Turku study[15] referred to above. Replacement of virtually all the sucrose in the diet of adult volunteers with xylitol for 2 years resulted in little development of new caries, and some early signs of demineralisation ('white spots') were apparently reversed or repaired (Table 4). By comparison, control subjects developed caries (cavities and pre-cavitation white spot lesions) in an average of 7·2 new tooth surfaces.

The lack of new caries in the xylitol group might be attributed to the non-fermentability of xylitol by the oral bacteria. However, the success of the sugar-substitution study prompted a further study[39] in which xylitol was administered only in one item (chewing gum) which was consumed approximately four times daily between meals, the remainder of the diet being unchanged. The results revealed a much lower caries incidence, despite the fact that only part of the dietary sugar was replaced by xylitol, perhaps indicating that the xylitol can prevent caries resulting from the sucrose in the rest of the diet, in addition to being incapable of causing caries itself.

Evidence of the lack of significant acid production from xylitol by plaque organisms is unequivocal, both from *in vitro* and *in vivo* studies;[3, 9,40-43] thus the *non-cariogenic* nature of xylitol is adequately explained. However, the suggestion of an *anti-cariogenic* action of xylitol has led to the search for evidence of inhibition of bacterial sucrose fermentation by xylitol, mostly with negative results.[42-45] A comprehensive microbiological experimental caries and human plaque study[46] did not show any inhibitory effects of xylitol on microbiological parameters; however, a non-significant reduction was observed in caries produced by feeding rats a diet containing sucrose when xylitol was added to the diet, but the animals consuming xylitol suffered from severe diarrhoea and gained less weight than the controls.

In the apparent absence of any inhibitory effects of xylitol on bacterial metabolism, other possible anti-caries effects might be accounted for by salivary mechanisms. Substitution of dietary sucrose with xylitol in human subjects and in monkeys led to an increase in the level of the enzyme lactoperoxidase in saliva. This enzyme catalyses the oxidation of salivary thiocyanate by hydrogen peroxide produced by some plaque organisms, leading to formation of a potent antibacterial substance, hypothiocyanite.[47] The significance of a rise in concentration of this enzyme is unknown—thiocyanate levels in the saliva did not alter, and it is not known if the enzyme activity or substrate availability limits the effectiveness of this antibacterial system. No increase in the level of lactoperoxidase was observed in the xylitol chewing gum experiment cited above[39] where an apparent anti-caries effect was observed.

The probable explanation of the caries-reducing effect of xylitol lies in the fact that its sweet taste stimulates a rapid flow of alkaline saliva together with a rise in salivary calcium levels, unaccompanied by any pH fall. These salivary changes might help to remineralise early damaged areas of enamel and tip the balance against the cariogenic action of the sucrose in the rest of the diet.[48]

Sorbitol and Other Polyols

Sorbitol is slowly fermented by oral organisms[2, 49, 50] and by plaque homogenates *in vitro*,[3, 43] but its effect on plaque pH *in vivo* is small.[29, 51, 52] Softening of experimental enamel slabs worn in the mouth and exposed to sorbitol is reduced compared to the effects of sucrose but is greater than with xylitol.[10] In rats and in monkeys, sorbitol diets have been found to be of low cariogenicity compared to glucose or sucrose diets.[53-55] Administration of sorbitol-containing between-meal sweets significantly reduced the development of caries over 3 years compared with controls consuming conventional sucrose-containing sweets according to a preliminary report on an experiment in institutionalised children in Hungary.[56]

Commercial hydrogenated starch hydrolysates, marketed under the trade name Lycasin®, contain a variable amount of sorbitol, maltitol and higher polyols.[57] Their fermentability by oral organisms is lower than sucrose and is further reduced if the maltitol content is high (50–55%).[3, 50] The high-maltitol products (and maltitol itself) do not give rise to a significant fall in plaque pH *in vivo*.[3] In a 2-year clinical study of the effect on caries in children of substituting sucrose in sweets with Lycasin, a reduction of approximately 25% was achieved, but this was of borderline significance and the results are complicated by a high rate of drop-out of subjects during the study and by availability of non-Lycasin containing sweets and beverages.

Caries experiments in hamsters and rats[58-60] have shown that Lycasin is less cariogenic than sucrose, but that the more hydrolysed product containing higher levels of maltitol was surprisingly more cariogenic than the less hydrolysed, more fermentable type. With the enamel slab technique, Lycasin and maltitol did not contribute to the softening of the enamel.[61]

A related sweetener (Palatinit) consisting of an equimolar mixture of [1→6] glucosyl-sorbitol and [1→6] glucosyl-mannitol has been found to give rise to little incidence of caries in rat experiments when compared with sucrose.[62-64] Mannitol itself is fermented only slowly by mixed plaque organisms *in vitro*,[43] and is less cariogenic than sucrose as judged by the softening of intra-oral enamel slabs and rat caries experiments.[10, 64] Lactitol resembles sorbitol in its fermentability.[50]

Other Sweeteners

Incubation of starch and sucrose with the enzyme cyclodextrin glycosyltransferase results in a mixture consisting mainly of glucosylsucrose and maltosylsucrose which is termed 'coupling sugar'. These sugars are fermented only slowly if at all by most strains of oral *Streptococcus mutans* tested,[65] and when fed to rats gave rise to little caries when the animals' mouths were mono-infected with this organism.[66] As a preliminary to further caries tests more work is required to determine whether coupling sugar is rapidly fermentable by the mixed human oral flora. Sorbose, a ketohexose produced by dehydrogenation of sorbitol, provokes only a small fall in pH in plaque *in vivo*[67] and has a low cariogenicity when applied five times daily to the teeth of rats.[64] This sweetener merits further attention.

Trichlorogalactosucrose (1', 4, 6'-trichlor-1', 4, 6'-trideoxygalactosucrose) has been found, in incubations with pure cultures of *Streptococcus mutans*, to possess metabolic inertness similar to that of xylitol.[42] It also has a small effect in reducing acid production from sucrose, perhaps by competing with the unsubstituted sugar for receptor sites for sugar translocation, and thus it might have an additional virtue in reducing caries produced by dietary sugar, as well as being itself non-cariogenic (Table 6).

TABLE 6
pH AFTER 2 H INCUBATION IN SALINE SUSPENSIONS OF *Streptococcus mutans* NCTC 10449 WITH VARIOUS SWEETENERS (DATA FROM REFERENCE 42)

Sweetener	pH
Sucrose	4·52
Glucose	4·53
Sucrose + xylitol	4·55
Sucrose + trichlorogalactosucrose	4·75
Sorbitol	4·81
Xylitol	5·32
Trichlorogalactosucrose	5·62

THE PLACE OF SUGAR ELIMINATION AND SUBSTITUTION IN CARIES PREVENTION

In seeking to effect a substantial change in caries incidence in the population through intelligent application of available dietary information, preventive dentistry must consider the emphasis to be placed on the use of less cariogenic sugar substitutes. Given the background to the caries problem, and the overwhelming evidence of the cariogenicity of sucrose, the case for the rational use of sugar substitutes is undeniable. However, the question to be decided, and one of interest to industry, is how best to employ these substitutes. Is it necessary to try to persuade people to abolish the use of sucrose altogether, and change to xylitol or sorbitol? Remembering that in western countries total sucrose consumption is not closely correlated with caries, but that the frequency of intake and nature of sugary foods—particularly between-meal sweets and snacks—is the important factor, then only part of the sugar we eat is cariogenic. The likelihood of mobilising the political, industrial and public opinion forces needed to achieve total substitution is exceedingly remote, but the elimination or substitution of the cariogenic forms of sucrose in the diet is more likely to be attainable.

In pursuing this aim, the first problem raised is how to identify these cariogenic forms. Since this impinges on the interests of commercial enterprises, it is necessary to be very specific about what the results of experiments to compare foods and beverages actually imply. Cariogenicity can only be proved in an absolute sense by clinical studies which may, for ethical as well as logistic reasons, be impossible. However, many manufacturers of foods which might be suspected of contributing to caries may see a commercial opportunity in incorporating some assessment of cariogenic potential in their product development programme.

The second problem arising from the attempt to use substitutes for the cariogenic part of our dietary sugar lies in the field of government regulation. Little advantage will accrue to food manufacturers who develop a product having apparently desirable characteristics from the dental point of view if they are not allowed to market it because of requirements of safety legislation, labelling regulations and codes of advertising practice. The requirements of government bodies for information on safety and efficacy are hampered by the complexity of the evidence in both spheres. The safety aspect is beyond the scope of this article but for the case of cariogenic potential, attempts have been and are being made to find an acceptable measure to provide a guide in product development and for use in labelling and advertising to inform consumers interested in modifying their diets to the benefit of their dental health.

Sugar substitution should not, however, be viewed in isolation. Dentists have advised the public for many years to avoid between-meal snacks and to reduce overall sugar consumption. This stern advice is unsympathetic to large sections of the public, particularly children, but is none the less valid. In addition, modification of the pattern of food consumption might assist in caries control—mention has been made above of the effect of eating sequence on the plaque pH response to a sugary food,[24] an idea supported by rat caries experiments using programmed sequences of sugar and non-sugary food.[68]

Protection of the teeth might also be achieved through the use of food additives. As caries is a process involving the dissolution of calcium phosphate salts, increasing the concentrations of calcium and/or phosphate in the plaque by adding suitable compounds to foods (especially those containing sugar) should, by the law of mass action, retard the caries process. Success has been achieved by the use of phosphate additives in animal experiments, but this approach has been less fruitful when applied to human populations.[69-72] Apart from these inorganic orthophosphate supplements, however, trimetaphosphate has been added to chewing gum and has been used with success to combat caries in children.[73] Organic phosphates may act by inhibiting enamel solubility rather than by providing a source of orthophosphate after hydrolysis by bacterial phosphatases. Calcium sucrose phosphate and calcium glycerophosphate have shown some promise and phytate (inositol hexaphosphate) is a powerful inhibitor of enamel solubility[74] and is effective in animal studies,[75, 76] although its use may be restricted by the reduction (perhaps temporary) in calcium absorption in the gut which is a well-known side effect of this compound.

This by no means complete catalogue of ways in which dietary modifications might be employed to combat caries should be viewed as part of the total programme on caries prevention, alongside various fluoride treatments including water fluoridation, and removal or inactivation of the bacteria which cause the disease, by mechanical, chemical or even perhaps by immunological means. The majority of caries is preventable; an important part of the attempt to reach this goal lies in assisting the public to correct its bad habits in the use of sugar.

REFERENCES

1. MARTHALER, T. M. (1978). In: *Health and Sugar Substitutes*, Ed. Guggenheim, B. Karger, Basel, 27–34.
2. EDGAR, W. M. (1976). *Caries Res.*, **10**, 241–54.
3. BIRKHED, D., and EDWARDSSON, S. (1978). In: *Health & Sugar Substitutes*, Ed. Guggenheim, B. Karger, Basel, 211–17.
4. KÖNIG, K. G., SCHMID, P. and SCHMID, R. (1968). *Archives oral Biol.*, **13**, 13–26.
5. BOWEN, W. H., AMSBAUGH, S. M., MONELL-TORRENS, S., COLE, M. and BRUNELLE, J. A. (1981). *Caries Res.*, **15**, 179 (Abstract).
6. FROSTELL, G. (1969). *Acta odont. Scand.*, **27**, 3–29.
7. EDGAR, W. M., BIBBY, B. G., MUNDORFF, S. and ROWLEY, J. (1975). *J. Am. dent. Ass.*, **90**, 418–25.
8. RUGG-GUNN, A. J., EDGAR, W. M. and JENKINS, G. N. (1978). *Brit. dent. J.*, **145**, 95–100.
9. IMFELD, T. (1977). *Schweiz. Mschr. Zahnheilk.*, **87**, 437–64.
10. KOULOURIDES, T., BODDEN, R., KELLER, S., MANSON-HING, L., LASTRA, J. and HOUSCH, T. (1976). *Caries Res.*, **10**, 427–41.
11. HARRIS, R. (1963). *J. dent. Res.*, **42**, 1387–99.
12. MARTHALER, T. M. (1967). *Caries Res.*, **1**, 222–38.
13. HOLLOWAY, P. J., JAMES, P. M. C. and SLACK, G. L. (1963). *Brit. dent. J.*, **115**, 19–25.
14. GUSTAFSSON, B., QUENSEL, C. E., LANKE, L., LUNDQVIST, S.,

GRAHNEN, H., BONOW, B. E. and KRASSE, B. (1953). *Acta odont. Scand.*, **11**, 232–363.
15. SCHEININ, A., MÄKINEN, K. K., and YLITALO, K. (1975). *Acta odont. Scand.*, **33**, Suppl. 70, 67–104.
16. GUGGENHEIM, B., KÖNIG, K. G., HERZOG, E. and MÜHLEMANN, H. R. (1966). *Helv. odont. Acta*, **10**, 101–13.
17. GREEN, R. M. and HARTLES, R. L. (1969). *Archives oral Biol.*, **14**, 235–41.
18. SHAW, J. H., KRUMINS, I. and GIBBONS, R. J. (1967). *Archives oral Biol.*, **12**, 755–68.
19. FROSTELL, G., KEYES, P. H. and LARSON, R. H. (1967). *J. Nutrition*, **73**, 65–7.
20. GRENBY, T. H. and HUTCHINSON, J. B. (1969). *Archives oral Biol.*, **14**, 373–80.
21. COLMAN, G., BOWEN, W. H. and COLE, M. F. (1977). *Brit. dent. J.*, **142**, 217–21.
22. FROSTELL, G., BLOMQVIST, T., BRUNER, P. O., DAHL, G. M., FJELLSTROM, A., HENRIKSON, C. O., LARJE, O., NORD, C-E., NORDENVALL, K, J, and WIK, O. (1981). *Caries Res.*, **15**, 200 (Abstract).
23. HUANG, C. T., LITTLE, M. F. and JOHNSON, R. (1981). *Caries Res.*, **15**, 54–9.
24. RUGG-GUNN, A. J., EDGAR, W. M., GEDDES, D. A. M. and JENKINS, G. N. (1975). *Brit. dent. J.*, **139**, 351–6.
25. HUXLEY, H. G. (1977). *Caries Res.*, **11**, 237–42.
26. KREITZMANN, S. N. and KLEIN, R. M. (1976). *J. dent. Res.*, **55**, Spec. Suppl. B, B175. (Abstract).
27. GREEN, R. M. and HARTLES, R. L. (1970). *Caries Res.*, **4**, 188–92.
28. HEFTI, A. and SCHMID, R. (1979). *Caries Res.*, **13**, 298–300.
29. FROSTELL, G. (1973). *Odont. Revy.*, **24**, 217–26.
30. JENKINS, G. N. and FERGUSON, D. B. (1966). *Brit. dent. J.*, **120**, 472.
31. WEISS, M. E. and BIBBY, B. G. (1966). *Archives oral Biol.*, **11**, 49–57.
32. KÖNIG, K. G. and GRENBY, T. H. (1965). *Archives oral Biol.*, **10**, 143–53.
33. KÖNIG, K. G. (1969). *Archives oral Biol.*, **14**, 991–3.
34. ISHII, T., KÖNIG, K. G. and MÜHLEMANN, H. R. (1968). *Helv. odnot. Acta*, **12**, 41–7.
35. BALEKJIAN, A. Y., TURNER, D. W., COTTON, W. R. and GUIDRY, M. S. (1980). *J. dent. Res.*, **59**, 2076–9.
36. MÜHLEMANN, H. R. and KÖNIG, K. G. (1967). *Helv. odont. Acta*, **11**, 152–6.
37. GRENBY, T. H. (1972). *Caries Res.*, **6**, 52–9.
38. GRENBY, T. H., POWELL, J. M. and GLESSON, M. J. (1974). *Archives oral Biol.*, **19**, 217–24.
39. SCHEININ, A., MÄKINEN, K. K., TAMMISALO, E. and REKOLA, M. (1975). *Acta odont. Scand.*, **33**, 269–78.
40. MÜHLEMANN, H. R. and DE BOEVER, J. (1970). In: *Dental Plaque*, Ed. McHugh, W. D. Livingstone, Edinburgh 179–86.

41. MÄKINEN, K. K. (1976). In: *Microbiological Aspects of Dental Caries*, Eds Stiles, H. M., Loesche, W. J. and O'Brien, T. Microbiol. Abs. Spec. Suppl. Information Retrieval Inc., Washington. 521–38.
42. DRUCKER, D. B. and VERRAN, J. (1979). *Archives oral Biol.*, **24**, 965–70.
43. HAYES, M. C. and ROBERTS, K. R. (1978). *Archives oral Biol.*, **23**, 445–51.
44. KLEBER, C. J., SCHIMMELE, R. G., PUTT, M. S. and MUHLER, J. C. (1979). *J. dent. Res.*, **58**, 614–8.
45. LUTZ, D. and GULZOW, H-J. (1979). *Caries Res.*, **13**, 132–6.
46. MÜHLEMANN, H. R., SCHMID, R., NOGUCHI, T., IMFELD, T. and HIRSCH, R. S. (1977). *Caries Res.*, **11**, 263–76.
47. HOOGENDOORN, H., PIESSENS, J. P., SCHOLTES, W. and STODDART, L. A. (1977). *Caries Res.*, **11**, 77–84.
48. LEACH, S. A. and GREEN, R. M. (1980). *Caries Res.* **14**, 16–23.
49. GUGGENHEIM, B. (1968). *Caries Res.*, **2**, 147–63.
50. HAVENAAR, R., HUIS IN'T VELD, J. H. J., BACKER DIRKS, O. and DE STOPPELAAR, J. D. (1979). In: *Health and Sugar Substitutes*, Ed. Guggengeim, B. Karger, Basel, 192–8.
51. FOSDICK, L. S., ENGLANDER, H. R., HOERMAN, K. C. and KESEL, R. G. (1957). *J. Am. Dent. Ass.*, **55**, 191–5.
52. BOWEN, W. H., EASTOE, J. E. and COCK, D. J. (1966). *Archives oral Biol.*, **11**, 833–7.
53. SHAW, J. H. and GRIFFITHS, D. (1960). *J. dent. Res.*, **37**, 377–84.
54. MÜHLEMANN, H. R., REGOLATI, B. and MARTHALER, T. M. (1970). *Helv. odont. Acta*, **14**, 48–50.
55. CORNICK, D. E. R. and BOWEN, W. H. (1972). *Archives oral Biol.*, **17**, 1637–48.
56. BANOCZY, J., HADAS, F. and ESZTÁRI, I. (1981). *Caries Res.*, **15**, 201 (Abstract).
57. FROSTELL, G. and BIRKHED, D. (1978). *Caries Res.*, **12**, 256–63.
58. FROSTELL, G. and BAER, P. N. (1971). *Acta odont. Scand.*, **29**, 253–9.
59. KARLE, E. and BÜTTNER, W. (1971). *Dt. Zahnarztl. Z.*, **26**, 1097–108.
60. BIRKHED, D. and FROSTELL, G. (1978). *Caries Res.*, **12**, 250–5.
61. RUNDEGREN, J., KOULOURIDES, T. and ERICSON, T. (1980). *Caries Res.*, **14**, 67–74.
62. KARLE, E. J. and GEHRING, F. (1978). *Dt. Zahnarztl. Z.*, **33**, 189–91.
63. VAN DER. HOEVEN, J. S. (1980). *Caries Res.*, **14**, 61–6.
64. FIRESTONE, A. R., SCHMID, R. and MÜHLEMANN H. R. (1980). *Caries Res.*, **13**, 324–32.
65. YAMADA, T., KIMURA, S. and IGARASHI, K. (1980). *Caries Res.*, **14**, 239–47.
66. IKEDA, T., SHIOTA, T., MCGHEE, J. R., OTAKE, S., MICHALEK, S. M., OCHIAI, K., HIRASAWA, S. and SUGIMOTO, K. (1978). *Infect. Immunity*, **19**, 477–80.
67. MÜHLEMANN, H. R. and SCHNEIDER, P. (1975). *Helv. odont. Acta*, **19**, 76–80.
68. EDGAR, W. M., BOWEN, W. H., AMSBAUGH, H. S. and MONELL-TORRENS, E. (1981). *Caries Res.*, **15**, 179 (Abstract).

69. NIZEL, A. E. and HARRIS, R. S. (1964). *J. dent. Res.*, **43**, 1123–36.
70. FINN, S. B. and JAMISON, H. C. (1967). *J. Am. dent. Ass.*, **74**, 987–95.
71. ASHLEY, F. P., NAYLOR, M. N. and EMSLIE, R. D. (1974). *Brit. dent. J.*, **136**, 361–6.
72. TATEVOSSIAN, A., EDGAR, W. M. and JENKINS, G. N. (1975). *Archives oral Biol.*, **20**, 617–25.
73. FINN, S. B., FREW, R. A., LEIBOWITZ, R., MORSE, W., MANSON-HING, L. and BRUNELLE, J. (1978). *J. Am. dent. Ass.*, **96**, 651–5.
74. JENKINS, G. N. (1966). *Adv. oral Biol.*, **2**, 67–100.
75. GRENBY, T. H. (1973). *J. dent. Res.*, **52**, 454–61.
76. COLE, M. F., EASTOE, J. E., CURTIS, M. A., KORTS, D. C. and BOWEN, W. H. (1980). *Caries Res.*, **14**, 1–15.

12

The Body Weight Response to Nutritional Sweeteners

I. MACDONALD
Department of Physiology,
Guy's Hospital Medical School,
London, UK

ABSTRACT

The majority of nutritional sweeteners in current use are di- or monosaccharides and there have been several isolated reports over the past 40 years to indicate that these 'simple' carbohydrates do not all have the same effect on body weight in experimental animals when given in equal amounts. There are also some preliminary studies in man that would appear to support the animal work

In the main these reports show that dietary sucrose is associated with more body weight gain than glucose when given as a high proportion of the energy intake, and in comparable amounts. It would seem, therefore, that the energy as determined in a calorimeter may not precisely reflect the metabolisable energy value of di- and monosaccharides, though the differences are probably only small in the context of the variations in energy intake in a person from day to day. However, a difference in body weight response to various 'simple' carbohydrates does exist, though its clinical significance is difficult to assess. The explanation for the difference may be in physical properties, endocrine response, biochemical pathways used, etc.

The majority of nutritional sweeteners currently of importance are di- and monosaccharides with, apparently, increasing use of partial hydrolysates of starch. It is, therefore, with these substances that most work has been carried out in assessing whether nutritional sweeteners all have equal effects on body weight when consumed in equal quantities.

It has been assumed in most studies that equal weights of carbohydrate sweeteners confer equal amounts of energy to the body; that is, that all 'simple' carbohydrates are iso-energetic gramme for gramme. Support for this assumption was received from experiments using calorimeters where the heat outputs from burning various 'simple' carbohydrates appeared to be almost identical. However, as will be seen, carbohydrate sweeteners that are iso-energetic in a calorimeter may not be iso-energetic in the body.

It was in 1935 that the first report appeared in which it was stated that rats given a high sucrose diet gained weight more rapidly than those given an iso-energetic glucose diet.[1] Thirty years later, again in rats, it was reported that animals with fructose or sucrose in their diet had a higher proportion of carcass fat than those consuming diets containing glucose.[2] By monitoring the rate of weight loss on a hypo-caloric intake it was found that rats lost weight more rapidly when the main energy source in the diet was glucose as compared with sucrose, despite the fact that both diets contained the same amount of energy as determined by the bomb calorimeter.[3]

One of the features of these hypo-caloric studies was that when, after a period of weight reduction, the diets were crossed over the rate of weight loss became equal in both groups of rats. Similarly, when the energy intake of rats which had been on a reducing regimen was increased, the rate of weight gain was identical for the two carbohydrates.[4] It thus seems possible that some 'adjustment' takes place which is dependent on the nature of the carbohydrate and that this alteration in metabolism persists for some considerable time. If confirmed, this would imply that the choice of carbohydrate sweetener at the beginning of a weight reducing regime could be important in determining the likelihood of a successful outcome.

Another species that shows differences in body weight between carbohydrate sweeteners when given equal amounts, is the baboon. The weight of the abdominal fat was significantly greater in baboons given sucrose than in those fed a partial hydrolysate of starch.[5] Baboons on a sucrose-containing diet gained more weight than those given partially hydrolysed starch.[6]

In a single, rather limited, study in man it was found that the body weight, in males, fell to a lesser extent on a reducing diet when the dietary carbohydrate was sucrose than when equal amounts of a partial hydrolysate of starch were given, while the reverse was seen in females.[7]

Thus it seems that the energy as determined in the laboratory using a calorimeter is not the same as the energy available to the body per unit weight of common carbohydrate sweetener and that the energy available to the body per gramme of carbohydrate varies according to the type of carbohydrate sweetener used. Though such differences exist, it is perhaps relevant to consider whether, if they occur in man, they are of any clinical significance. The energy intake of the body is remarkably well controlled by a mechanism or mechanisms not understood and in the majority of people the error of these mechanisms is minute. It seems likely, therefore, that metabolisable energy differences between carbohydrates, such as those seen in experimental animals, might well be compensated for in man whose adult weight is constant—unlike the rat and baboon whose adult weight increases physiologically. On the other hand there are many individuals whose energy control system is either overridden or has a wide margin of error, in which case small differences between the available energy of carbohydrate sweeteners could, especially if consumed in large amounts, lead to disproportionate changes in body weight, even when food is consumed in equal quantities.

All this is speculative and is likely to remain so until the biological explanation for the difference in metabolisable energy between carbohydrate sweeteners is discovered. There are several possible explanations and some of these will be considered.

1. Sucrose has to be hydrolysed before its constituent monosaccharides can be absorbed and in the process of hydrolysis heat is produced. The theoretical aspects have been considered[8] and do undoubtedly contribute to the finding that, weight for weight, sucrose seems to be used by the body more efficiently than glucose or its polymers. The extent of this heat of hydrolysis can be studied in rats by comparing sucrose against its constituent monosaccharides, glucose and fructose, in the diet, but in man this would be difficult in view of the diarrhoea liable to occur after sizeable intakes of fructose.

2. Related in part to the heat of hydrolysis is the extent of the increase in metabolic rate (dietary-induced thermogenesis)

following ingestion of the carbohydrate sweeteners. When fasting rats are given sucrose the rise in metabolic rate is more rapid than with glucose, though this initial rise with sucrose falls sooner than with glucose. Glucose and fructose produce similar increases in metabolic rate during the 3 h after ingestion. It is perhaps of significance, that the metabolic rate in rats is significantly greater after sucrose ingestion than after ingestion of an equivalent mixture of its two constituent monosaccharides.[9] In 1918 Benedict and Carpenter[10] reported that, in man, the increment in heat production was much lower with glucose than with sucrose, and similar results have been reported in a dog.[11] These findings in man have been confirmed more recently with similar differences between glucose and sucrose seen in overweight persons.[12] All these findings would not support the view that sucrose is used more 'efficiently' by the body, in fact the reverse, but the studies above were only carried out for periods of 3 h or so and perhaps over a longer period of time the overall increase in metabolic rate might have been lower for sucrose.

3. Despite the evidence that a smaller increase in metabolic rate is not seen after sucrose ingestion compared to glucose ingestion in the short term as would be expected, the overall increment might be less. If this is so then there are pointers which would support the view that the biochemical processing of the carbohydrate sweeteners varies. For instance the respiratory quotient is significantly greater following the ingestion of sucrose than following that of glucose both in man and rats,[9,12] and in man the acute ingestion of sucrose and fructose is followed by an increase in pyruvate, lactate and uric acid whereas this is not found after glucose ingestion.[13] These acute differences could be explained by the less rapid metabolism of glucose compared to fructose,[14] the hexokinase involved in the early stages of glucose metabolism being rate limiting[15] whereas fructo-hexokinase, which catalyses fructose phosphorylation, is not.[16] Though no precise explanation for the differences in body weight response between carbohydrate sweeteners is yet apparent, sufficient differences are known in their biochemical handling not to exclude this as being possibly responsible.

4. Certain hormones are known to affect overall metabolism and

metabolic rate, the two obvious ones being insulin and thyroid. The insulin response to fructose is very small and probably reflects that part of the fructose that has been converted to glucose,[17,13] such that the insulin response to sucrose is half that to an equal amount of glucose. The fact that the insulin response to oral ingestion of glucose is much greater than that to the same amount given intravenously[18] would suggest that a role in metabolism played by the hormones produced in the gut wall cannot be excluded. More recently it has been suggested that, in man, the output of hormones from the thyroid might vary, in the short term, for sucrose and glucose.[19]

Thus various effects are noted after the ingestion of different carbohydrate sweeteners but which, if any, of these is responsible for the difference in body weight between the various carbohydrate sweeteners investigated remains to be seen.

REFERENCES

1. FEYDER, S. (1935). *J. Nutr.*, **9**, 457–68.
2. ALLEN, R. J. L. and LEAHY, J. S. (1966). *Br. J. Nutr.*, **20**, 339–47.
3. MACDONALD, I. and GRENBY, T. H. (1979). *Proc. Nutr. Soc.*, **38**, 30A.
4. MACDONALD, I., GRENBY, T. H., WILLIAMS, C. and FISHER, M. (1981). *J. Nutr.* (in press).
5. ALLEN, R. J. L., BROOK, M., LISTER, R. E. and SIM, A. K. (1966). *Nature*, **211**, 1104.
6. BROOK, M. and NOEL, P. (1969). *Nature*, **222**, 562–3
7. MACDONALD, I. and TAYLOR, J. (1973) *Guy's Hosp. Rep.*, **122**, 155–9.
8. JACKSON, R. J. and DAVIS, W. B. (1977). *Proc. Nutr. Soc.*, **36**, 90A.
9. SHARIEF, N. Y. and MACDONALD, I. (1979). *Proc. Nutr. Soc.*, **38**, 83A.
10. BENEDICT, F. G. and CARPENTER, T. M. (1918). *Carnegie Ins. Wash. Public*, **261**, 47–250.
11. LUSK, G. (1915). *J. Biol. Chem.*, **20**, 555–617.
12. SHARIEF, N. Y. and MACDONALD, I. (1980). *Proc. Nutr. Soc.*, **39**, 42A.
13. MACDONALD, I., KEYSER, A. and PACY, D. (1976). *Am. J. Clin. Nutr.*, **31**, 1305–11.
14. BERGSTROM, J. and HULTMAN, E. (1967). *Acta Med. Scand.*, **182**, 93–107.
15. HEINZ, F. (1973). In: *Progress in Biochemical Pharmacology*, Vol. 8, Ed. Macdonald, I. Karger, Basel. 1–56.
16. SMITH, L. H., ETTINGER, R. H. and SELIGSON, D. (1953). *J. Clin. Invest.*, **32**, 273–82.

17. ROZEN, P. and SHAFRIR, E. (1972). *Israel J. Med. Sci.*, **8**, 838–40.
18. MCINTYRE, N., HOLDSWORTH, C. D. and TURNER, D. S. (1964). *Lancet*, **2**, 20–1.
19. SHARIEF, N. Y. and MARSDEN, P. (1981). *Proc. Nutr. Soc.*, **40**, 43A.

13

Obesity, Thermogenesis and Carbohydrate Metabolism

M. J. STOCK and NANCY J. ROTHWELL
*Department of Physiology,
St. George's Hospital Medical School,
Tooting, London, UK*

ABSTRACT

Human obesity is usually ascribed to an inability to control food intake, even though many studies have failed to detect higher energy intakes in obese people compared to lecn. The results of several overfeeding studies suggest that lean subjects resist weight gain by increasing metabolic rate in response to increases in energy intake.

These experiments have now been simulated in rats by feeding them a large variety of palatable food items. This 'cafeteria' diet produces very large (80%) increases in voluntary food intake. However, depending on the age and strain of the rats, this overeating does not necessarily produce obesity. In those animals capable of resisting excess weight gain there are compensatory increases in heat production and this 'diet-induced thermogenesis' (DIT) is mainly due to increased activity of the sympathetic nervous system. Changes in mitochondrial function, responses to sympathetic stimulation and in vivo rates of oxygen consumption implicate brown adipose tissue as the major source of DIT. In spite of large increases in carbohydrate intake on the cafeteria diet, blood glucose levels remain normal whilst plasma insulin levels are lower than normal. This suggests that, in animals capable of resisting obesity, increases in food intake result in improved glucose homeostasis rather than the hyperglycaemia and hyperinsulinaemia seen in obesity. The role of insulin in DIT may be complex since studies with diabetic animals indicate an insulin requirement for thermogenesis and more recent work has also implicated an insulin-sensitive area of the hypothalamus in the sympathetic activation of brown fat thermogenesis. Insulin may therefore act as a central signal for activating DIT as well as exerting its permissive effects on peripheral brown fat metabolism.

INTRODUCTION

Obesity is a condition resulting from excessive triglyceride storage and yet some of the most noticeable associated metabolic changes relate to carbohydrate metabolism. The association between obesity and diabetes in man has been recognised since antiquity and is a prominent feature of the genetic obesity seen in laboratory rodents such as ob/ob and db/db mice and the fa/fa rat.[1] The investigation of this association between excess lipid storage and impaired carbohydrate metabolism has occupied many workers, and the causal links have proved particularly difficult to resolve. The genetically obese rodents usually exhibit hyperinsulinaemia and insulin resistance but in some (e.g. db/db) this can appear before obvious signs of obesity develop, while in others (e.g. ob/ob and fa/fa) obesity can be detected before the changes in pancreatic function. In man, it would appear that the non-insulin-dependent diabetes prevalent in middle-aged Western society is precipitated by obesity but the fact that not all obese individuals are affected suggests that excess fat deposition could be unmasking a genetically determined trait.

The intimate association between obesity and diabetes has inevitably led to the suggestion that both are the consequence of overnutrition, particularly when a large fraction of energy is derived from refined dietary carbohydrates.[2,3] There can be no doubt that obesity is due to energy intake exceeding expenditure but this does not necessarily result from gross aberrations in appetite and high levels of food consumption. It is becoming increasingly obvious that obesity involves a metabolic derangement which affects the ability to dispose of excess energy via increases in heat production (i.e. thermogenesis). In this review the evidence for this energy dissipating mechanism and its relevance to the aetiology of obesity will be considered. Preliminary data on the effects of carbohydrate and insulin on thermogenesis will be presented and the relationship between thermogenesis and insulin as affected by age, environment and genetics will also be discussed.

THERMOGENESIS IN MAN

The idea that changes in the efficiency of food utilisation could contribute to the regulation of energy balance is by no means new. At the turn of this century Rubner[4] and Neumann[5] suggested just such a mechanism which Neumann called 'luxoskonsumption'. However, it was not until the mid-1960s that determined efforts were made to investigate this area of metabolism. Three separate human overfeeding studies were carried out by groups in London,[6] Paris[7] and Vermont[8] and in each study there was evidence for adaptive changes in metabolism induced by increases in energy intake.

Although there were differences between each study, as well as individual differences in the response to overnutrition, it was evident that many of the subjects had a capacity to resist excessive weight gain. Surprisingly, even those subjects who did gain a large amount of extra weight found it difficult to maintain this and the Vermont group found that these subjects required twice as much energy for weight maintenance as non-experimental, spontaneously obese subjects.[8] In the three overfeeding studies there was evidence (either direct or indirect) for increases in metabolic rate and this effect of food intake on heat production is now known as 'diet-induced thermogenesis' (DIT), rather than the earlier term of luxoskonsumption.

The ability of some people to resist weight gain by increasing DIT suggests that obesity could be due to a failure to activate this form of heat production following the consumption of food. There is certainly some evidence for this since obese women exhibit a reduced thermogenic response to a standard meal compared to lean women.[9] A more remarkable demonstration of individual differences in the response to food is provided by the study of York et al.[10] These workers measured the metabolic responses of a group who, although having similar body weights, could be divided into 'large' eaters (customary intake, 16·8 MJ/day) and 'small' eaters (7·1 MJ/day). The increase in metabolic rate (i.e. DIT) following a 2·1 MJ meal was 29% and 7% in the 'large' and 'small' eaters respectively, thus suggesting that the 'small' eaters were more efficient and therefore more prone to obesity than the 'large' eaters. This paradoxical conclusion appears to run completely contrary to conventional views on the relationship between food intake and weight gain but is justified by the observation that, in spite of similar body weights, the 'small' eaters had a higher body fat content (22·5%) than the 'large' eaters (12·6%).

If one accepts the results of these various studies on the variability of metabolic efficiency, it should come as no surprise to find a 2–3-fold range of food intake in populations of subjects of the same age and weight,[11,12] or to find that obese people do not necessarily eat more than lean people.[13-15] These findings are hard to reconcile with the concept of gluttony as the cause of obesity and over the past 10 years there has been a steady shift towards the view that defective or inadequate DIT is a major contributory factor to the development of obesity. Much of the supporting evidence for this and for an active role of DIT in energy balance regulation comes from animal studies.

CAFETERIA FEEDING AND THERMOGENESIS IN RATS

Much of the early evidence for adaptive thermogenesis in animal nutrition came from studies using low protein diets,[16-19] but the recent introduction of 'cafeteria' feeding[20,21] has meant that the effects of voluntary hyperphagia on DIT can be observed in rats eating nutritionally adequate diets. The effects of presenting rats with a palatable and varied diet, consisting of foods usually eaten by man, is to increase energy intake by as much as 80% above that of rats fed standard laboratory diets. Apart from eating a greater bulk of food, the higher energy intake of cafeteria-fed rats is due to an increase in the energy density (fat accounts for 35–40% of energy) and digestibility of the diet. Rats seem to choose a diet with a nutrient composition very similar to that eaten by 'supermarket' man, with protein accounting for 15–16% of energy in young rats, falling to 11–12% as they grow older. This feeding system therefore provides an ideal method for simulating human feeding practices in laboratory animals. It also demonstrates that the rat is as susceptible as man to the hedonistic qualities of food and that control of energy intake in the rat is not the precise physiological function we have been led to believe.

The changes in body weight and fat content of cafeteria-fed rats appear to be as variable as those seen in the human studies discussed above. Depending on age and strain (and probably other factors yet to be investigated) some rats readily become obese,[21] while others either gain no extra weight[22] or gain weight with difficulty.[23] This reduction in efficiency of weight gain in cafeteria-fed rats has been observed by other workers[24] and the authors of this paper have demonstrated that it is due to increases in heat production (i.e. DIT) that can be as much as 100% above control values.[23] Thus, by choosing an appropriate age and strain of rat, one can produce an animal that overeats by 80% but stays lean.

NON-SHIVERING AND DIET-INDUCED THERMOGENESIS

The cafeteria-fed rat exhibiting DIT resembles the cold-adapted rat, which also has a high food intake and an elevated rate of heat production. The greater thermogenesis of cold-adapted rats is not due to shivering (i.e. muscle thermogenesis) but to a non-shivering thermogenesis (NST) mainly resulting from sympathetic activation of brown adipose tissue (BAT) metabolism. Brown adipose tissue, or brown fat, is a richly innervated and well-vascularised form of adipose tissue with much less triglyceride than white fat but containing numerous mitochondria, which provide it with a large capacity for oxidation of substrates. It is usually found around major organs (heart and kidney) and in superficial sites, such as between the scapulae and in the axilla. Its total mass is small ($<1\%$ of body weight) but, because it has a good blood supply and high respiratory capacity, BAT can account for over 60% of NST.[25]

The superficial resemblance between DIT and NST prompted further investigation and it was found that the metabolic changes induced by cafeteria feeding were practically identical, qualitatively and quantitatively, to those seen in cold adaption. The similarities include: increased cold tolerance in the absence of shivering; a doubling of the capacity to respond to the thermogenic effects of noradrenaline; inhibition of DIT by β-adrenergic blockade; and hypertrophy and hyperplasia of BAT.[23, 24, 26]

The biochemical origins of brown fat thermogenesis have yet to be elucidated fully but both a proton conductance pathway in BAT mitochondria[27] and increased Na^+K^+-ATPase activity[28] have been implicated. Both of these thermogenic mechanisms show increased activity in BAT obtained from cafeteria-fed rats exhibiting

DIT[29, 30] and provide strong support for the existence of a role for BAT in DIT. However, the most convincing evidence comes from *in vivo* measurements of BAT oxygen consumption, which demonstrate that all of the diet-induced changes in thermogenic capacity result from increases in BAT thermogenesis.[31]

The results of these studies with cafeteria feeding have firmly established DIT as a quantitatively important mechanism in energy metabolism and have placed a new role and emphasis on the sympathetic control of BAT. Until recently, this tissue had been largely ignored by most physiologists, apart from those interested in the thermoregulation of animals exposed to the cold. However, at the same time as the role of BAT in cafeteria-fed rats was being investigated, a group at the MRC Dunn Nutrition Unit (Cambridge) was pursuing the problem of defective thermogenesis exhibited by genetically obese animals. In a remarkable demonstration of how a new idea can emerge from opposite ends of a problem, this group arrived at very similar conclusions regarding the role of BAT in the regulation of energy balance.

NST, DIT, BAT AND OBESITY

The genetically obese rodents (e.g. ob/ob and db/db mice and fa/fa rat) generally exhibit hyperphagia, but this does not entirely account for their obesity. Careful pair-feeding studies have shown that equal energy intakes result in greater rates of weight gain in obese animals than their lean littermates.[32-34] This implicates a greater metabolic efficiency (i.e. lower energy expenditure) as a contributory factor in the development of obesity. Measurements of oxygen consumption can in fact be used to detect the obese genotype before the overt phenotypic signs of obesity develop.[35] An alternative method for early detection of the obese genotype relies on the fact that these animals fail to maintain body temperature during cold exposure and can die of hypothermia.[36] Pair-feeding studies over a range of environmental temperatures have also shown that the greater energy retention of ob/ob mice is more marked at low temperatures, although it is still evident at thermoneutral (i.e. warm) temperatures.[33]

The results of these experiments indicate that the impairment in thermogenesis is a contributory factor in the development of obesity and subsequent studies have demonstrated that this defect involves the same mechanisms as those responsible for NST and DIT. For example, the capacity of obese animals to respond to noradrenaline is reduced[37,38] and this is due to the lower activity of BAT.[39] This defect in BAT thermogenesis is probably due to the fact that the proton conductance pathway of BAT mitochondria is poorly developed.[40,41] The investigation of these mechanisms in human obesity has hardly begun but there is evidence that similar metabolic changes occur. Lean subjects exhibit better tolerance and higher metabolic rates than obese subjects during cold exposure[42] and the thermogenic response of obese women infused with noradrenaline is less than that of normal-weight women.[43] Evidence for NST in man is gradually accumulating[44] and the presence of BAT in adult man,[45,46] which may have retained its thermogenic capacity,[23] indicates that the animal studies on thermogenesis could have a direct bearing on the aetiology of human obesity.

At the beginning of this review the close relationship that exists between obesity and diabetes was mentioned. In view of the evidence for a thermogenic defect in obesity, it is logical to consider possible relationships between carbohydrate metabolism and thermogenesis, particularly in situations where carbohydrate intake is high, e.g. in the cafeteria-fed rat.

A ROLE FOR INSULIN IN DIT

The carbohydrate content of most laboratory rat diets is high (approximately 75% by weight) but much of this is fibre and the digestibility of these diets is consequently very low (60–70%). This has to be contrasted with the cafeteria diet of rats fed human foods where carbohydrate accounts for 50% of nutrient intake but, because of its high digestibility (90–95%) and the large increases in total food intake, the energy derived from carbohydrate is nearly twice that of stock-fed animals. According to some authorities this high intake of refined carbohydrate should produce some of the dire metabolic consequences (e.g. obesity and diabetes) seen in human populations eating low fibre, high energy diets. However, as discussed above, cafeteria feeding does not necessarily produce obesity in rats and it has now been found[47] that neither is glucose homeostasis impaired. Instead of hyperglycaemia and hyperinsulinaemia, normoglycaemia, concomitant with reductions in circulating insulin, has been observed, which suggests that improved glucose utilisation and insulin responsiveness accompany DIT.

Many nutritionists may find this amelioration of glucose control during overnutrition difficult to accept, but a similar response is seen during cold exposure[48] when animals are exhibiting NST. In both NST and DIT the activity of the sympathetic nervous system is high and it is possible that this could cause a reduction in insulin levels[49] as well as exerting a stimulatory effect on glucose uptake.[50] The changes in insulin levels in hyperphagic rats exhibiting DIT prompted the authors to investigate the role of this hormone in thermogenesis using diabetic rats.[47]

Diabetes was induced by streptozotocin but, to avoid the gross metabolic effects of insulin deficiency, the animals were given replacement doses of a long-acting insulin preparation (protamine zinc insulin, PZI) every 48 h. This replacement schedule resulted in chronic responses to cafeteria feeding that were essentially the same as in non-diabetic animals. However, towards the end of each 48 h period (i.e. 36 h after the last dose of PZI) the typical changes in metabolism seen in cafeteria rats were absent. For example, resting metabolic rates were identical to those of stock-fed controls, indicating low levels of DIT, and the thermogenic response to noradrenaline was also similar (Fig. 1). These measurements were made at a time when the animals were deficient in insulin and quite different results were obtained when the measurements were repeated just 12 h after the PZI injection (Fig. 2). At this time, the diabetic cafeteria rats exhibited increases in resting metabolic rate and thermogenic responses to noradrenaline that were indistinguishable from those of non-diabetic cafeteria rats. Thus, it would seem that as long as insulin is present, the effects of cafeteria feeding on energy dissipation are evident but these effects disappear during acute insulin deficiency.

The results of these experiments have led to the suggestion that insulin plays a permissive role in the development of DIT and this could be partly due to the effects of insulin on BAT metabolism. Preliminary *in vitro* studies provide some evidence that insulin may facilitate the effects of noradrenaline on BAT heat production and lipolysis[51] but as well as these potential peripheral effects of insulin, a central role in thermoregulation and energy balance is also indicated. In the diabetic study described above it was found that

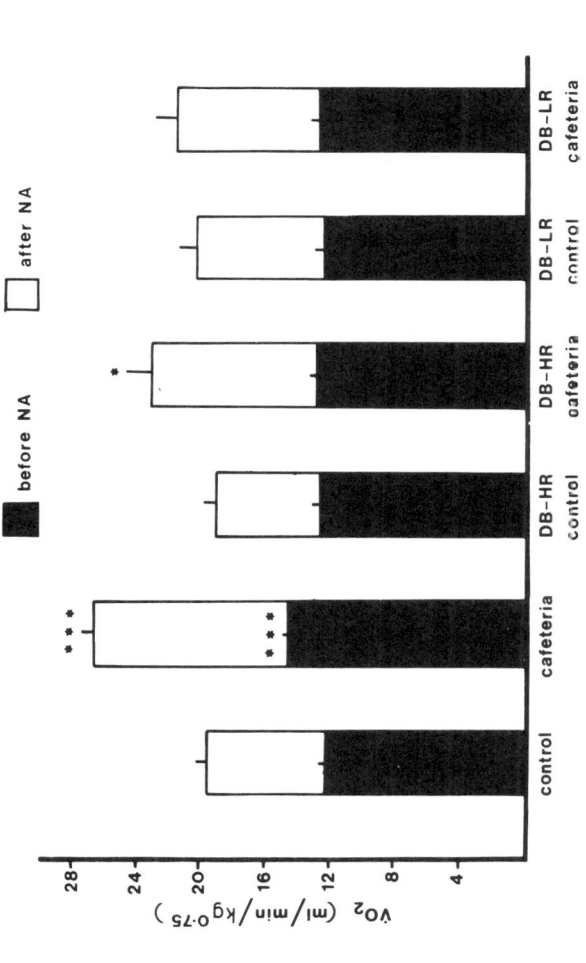

FIG. 1. Resting oxygen consumption ($\dot{V}O_2$, expressed per unit metabolic body size, $kg^{0.75}$) before and after noradrenaline (25 μg/100 g body weight, s.c.) in normal and diabetic rats fed stock diet (control) or the cafeteria diet. DB–HR and DB–LR refer, respectively, to diabetic groups receiving 4 units or 2 units of PZI every 48 h. The measurements of oxygen consumption shown were made 36 h after the last dose of PZI. (Mean values + SEM; $n=8$.) $*p<0.05$; $***p<0.001$ compared to respective control.

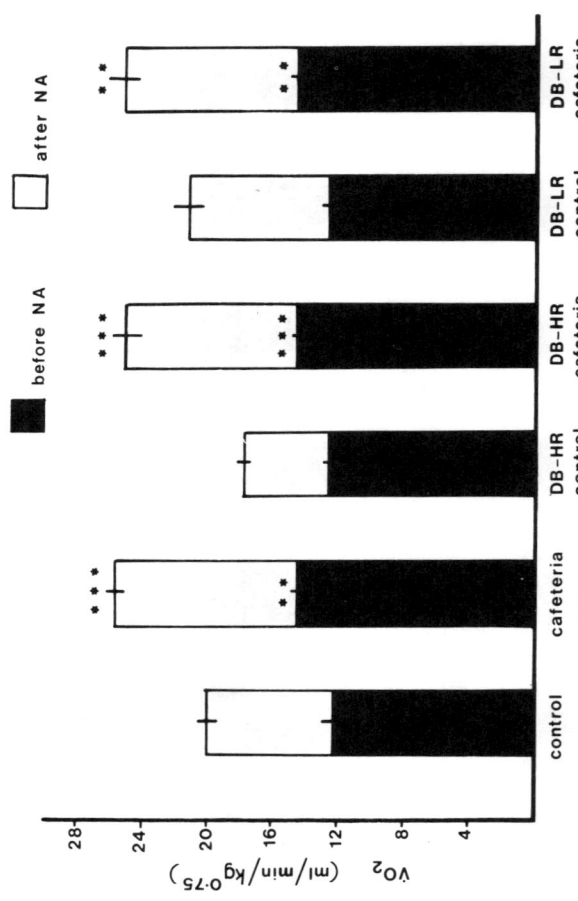

FIG. 2. The measurements shown in this figure are the same as for Fig. 1 except that they were made 12 h after the last dose of PZI.

the thermogenic responses to noradrenaline in control and cafeteria rats during acute insulin deficiency were practically identical to responses in non-diabetic stock-fed animals. However, during cold exposure (2 h at 5°C) all the insulin-deficient rats became markedly hypothermic. This hypothermia must presumably result from a failure of hypothalamic thermoregulatory mechanisms to activate thermogenesis, since the effector tissues were capable of responding to exogenous noradrenaline.[52] Other workers[51] have also noted hypothermia in alloxan diabetic rats and found that insulin replacement was capable of reversing this impairment in thermogenesis.

These studies provide another example of changes in NST accompanying changes in DIT, but also suggest that insulin may be required for central activation of thermogenesis. It has been known for some time that certain areas of the brain, particularly the ventromedial hypothalamus (VMH), are insulin-sensitive and the VMH is also known to exert a profound influence on food intake. It has now been demonstrated that electrical stimulation of this same area results in activation of BAT thermogenesis[53] and Seydoux et al.[54] have shown that VMH lesions inactivate BAT. It is therefore plausible, but by no means proven, that diet-induced changes in thermogenesis are mediated by insulin acting via the hypothalamus. Landsberg et al.[55] have also argued that the increases and decreases in sympathetic activity that respectively accompany overfeeding and fasting are probably mediated by changes in circulating glucose or, more likely, by changes in insulin. Rises in plasma noradrenaline (in the absence of hypoglycaemia) have been noted following insulin infusions[56] and this observation provides further support for the idea that insulin could act as the humoral signal for activation of thermogenesis.

CARBOHYDRATES AND DIT

The interactions between insulin and thermogenesis discussed above suggest that the carbohydrate content of the diet could be an important determinant of metabolic efficiency. It is known that carbohydrate intake influences other hormones, apart from insulin, and these could also affect energy metabolism. Thyroid hormones are a particular example and, although the precise role of these hormones in DIT is uncertain and beyond the scope of this review, some of the dietary responses are worth describing. Overnutrition in man[8] and rats[23] produces increases in plasma triiodothyronine (T_3) levels, probably as a result of increased peripheral conversion of thyroxine to T_3. This effect can also be observed without any increase in food intake if carbohydrate is substituted for fat in the diet.[57] Conversely, starvation is associated with decreases in T_3 which can be simulated without fasting by substituting fat for carbohydrate.[5] Although starvation results in decreased metabolic rates, these can be dissociated from the changes in thyroid hormone metabolism because they still occur in thyroidectomised animals as well as in thyroidectomised animals receiving constant replacement doses of T_3.[58]

When fasted rats are given a small (40 kJ) carbohydrate meal, there is a transient rise in T_3 levels followed 12 h later by a rise in metabolic rate which reaches a peak 24 h after the meal. A meal of fat or an injection of T_3 fail to produce this rise in metabolic rate, although a combination of the two does.[59] It has since been found that both the rise in T_3 and the subsequent increase in metabolic rate produced by the carbohydrate meal can be blocked by the β-adrenergic antagonist propranolol.[60] This indicates a sympathetic involvement in the diet-mediated changes in thyroid hormone metabolism as well as in the thyroid-dependent changes in heat production, and illustrates the complex relationships that exist between dietary carbohydrate, the thyroid and the sympathetic nervous system. The results also suggest that the depression of metabolism seen when obese subjects are starved might be avoided by giving very small, frequent carbohydrate meals, since these would serve to maintain T_3 levels and sympathetic tone which might result in more consistent weight losses.

In experiments utilising cafeteria feeding, the extent to which the carbohydrate content of the diet determines the degree of hyperphagia and levels of DIT is unknown. Recently, however, studies have been started to investigate this by feeding high carbohydrate, low fat (HC) and high fat, low carbohydrate (HF) cafeteria diets. Cafeteria feeding obviously requires a variety of foods to induce voluntary hyperphagia and the major difficulty has been to find foods with an appropriate carbohydrate:fat ratio, whilst ensuring an adequate protein content. In a preliminary trial hyperphagia was successfully produced with intakes of fat, carbohydrate and protein (all expressed as percentage total energy intake) of 5, 81 and 14%(HC) and 84, 3 and 14%(HF), respectively.

Animals on the HF cafeteria diet more than doubled their energy intake whereas HC cafeteria rats increased their intake by only 47% relative to stock-fed controls. This suggests that hyperphagia on high carbohydrate diets is limited by bulk, because of the low energy density of the food items, although the low fat content may also result in a less palatable diet. The rats used for this study were adult and on the HF diet gained more weight than stock-fed controls, although the excess weight gain was small due to adaptive increases in DIT. Surprisingly, rats on the HC cafeteria diet gained no excess weight even though energy intake was elevated by 47%. These results are preliminary and should be treated with caution but they imply that high carbohydrate/low fat diets might be less likely to produce obesity because (i) the degree of hyperphagia could be limited by energy density and palatability, and (ii) DIT is more pronounced.

BIOLOGICAL IMPLICATIONS

The reasons why leanness and thermogenesis could have a positive selective advantage in the evolution of mammals have been set out in an earlier paper by the authors;[61] other workers[62,63] have argued that on the other hand thrifty genes associated with energy conservation, obesity and diabetic tendencies could also have selective value. The latter features would certainly help animals to survive in hot, arid desert regions and this might explain why animals (e.g. spiny mouse, sand rat and the tuco tuco) and men (e.g. Aborigines and the Pima Indians) from such areas readily become obese and/or diabetic when provided with adequate food. However, in more temperate regions these thrifty genotypes are at a disadvantage, not only because frank obesity and diabetes affect growth, fertility, mobility, etc., but because the associated defects in thermogenesis make them more prone to hypothermia. The susceptibility of genetically obese rodents to cold exposure illustrates this, as does the observation[64] that diabetic patients accounted for 20% of admissions for accidental hypothermia.

The evolutionary interactions between obese/diabetic genotypes and the environment could be used to explain ethnic variations in the prevalence of diabetes.[65] For example, the prevalence rate of diabetes is about 19% in urbanised Aborigines[66] and exceeds 30% in Pima Indians[67] but is less than 0·5% in Eskimos.[68] Provided overt diabetes is not expressed, it is possible that many populations have retained carriers of the thrifty genotype, particularly in civilised societies where the biological pressures on survival have been minimised. Thus, one could postulate that the development of obesity and diabetes requires two factors—a genetic predisposition and adequate food. Neither factor alone is sufficient, and those who do not possess thrifty genes can indulge their appetite without fear of the consequences, although they are likely to be the first victims of famine if they cannot control their thermogenesis.

REFERENCES

1. BRAY, G. A. and YORK, D. A. (1979). *Physiol. Rev.*, **59**, 719–809.
2. COHEN, A. M., TEITELBAUM, A. and SALITERNIK, R. (1972). *Metabolism*, **21**, 235–41.
3. CLEAVE, T. L. (1974). *The Saccharine Disease*, John Wright, Bristol.
4. RUBNER, M. (1902). *Die Gesetze des Energieverbrauchs beider Ernahrung*, Deutiche, Leipzig.
5. NEUMANN, R. O. (1902). *Arch. Hyg.* **45**, 1–87.
6. MILLER, D. S. and MUMFORD, P. (1967). *Am. J. Clin. Nutr.*, **20**, 1212–22.
7. APFELBAUM, M. (1974). In: *Obesity in Perspective*, Ed. Bray, G. A. DHEW Pub. No. (NIH) 75–708, Washington, DC, 127–36.
8. SIMS, E. A. H., DANFORTH, E., HORTON, E. S., BRAY, G. A., GLENNON, J. A. and SALANS, L. B. (1973). *Rec. Prog. Horm. Res.*, **29**, 457–96.
9. KAPLAN, M. L. and LEVEILLE, G. A. (1976). *Am. J. Clin. Nutr.*, **29**, 1108–13.
10. YORK, D. A., MORGAN, J. B. and TAYLOR, T. G. (1980). *Proc. Nutr. Soc.*, **39**, 57A.
11. WALKER, S. E. (1965). *Br. J. Nutr.*, **19**, 1–12.
12. WIDDOWSON, E. M. (1947). MRC Spec. Rep. Ser. No. 257, London.
13. JOHNSON, M. L., BURKE, B. S. and MAYER, J. (1956). *Am. J. Clin. Nutr.*, **4**, 231–8.
14. MCCARTHY, M. C. (1966). *J. Am. Diet. Ass.*, **29**, 29–33.
15. STEFANIK, P. A., HEALD, F. P. and MAJER, (1959). *Am. J. Clin. Nutr.*, **7**, 55–62.
16. MILLER, D. S. and PAYNE, P. R. (1962). *J. Nutr.*, **78**, 255–62.
17. STIRLING, J. L. and STOCK, M. J. (1968). *Nature*, **220**, 801–2.
18. MCCRACKEN, K. J. and GRAY, R. (1976). *Proc. Nutr. Soc.*, **35**, 59A–60A.

19. GURR, M. I., MAWSON, R., ROTHWELL, N. J. and STOCK, M. J. (1980). *J. Nutr.*, **110**, 532–42.
20. SCLAFANI, A. and SPRINGER, D. (1976). *Physiol. Behav.*, **17**, 461–71.
21. ROTHWELL, N. J. and STOCK, M. J. (1979) *J. Comp. Physiol. Psychol.*, **93**, 1024–34.
22. ROTHWELL, N. J. and STOCK, M. J. (1980). *Proc. Nutr Soc.*, **39**, 5A.
23. ROTHWELL, N. J. and STOCK, M. J. (1979) *Nature*, 281, 31–5.
24. TULP, O., FRINK, R., SIMS, E. A. H. and DANFORTH, E. (1980). *Clin. Res.*, **28**, 621A.
25. FOSTER, D. O. and FRYDMAN, M. L. (1978). *Can. J. Physiol. Pharmacol.*, **56**, 110–22.
26. ROTHWELL, N. J. and STOCK, M. J. (1980). *Can. J. Physiol. Pharmacol.*, **58**, 842–48.
27. NICHOLLS, D. G. (1979). *Biochem. Biophys. Acta.*, **549**, 1–29.
28. HORWITZ, B. A. (1973). *Am. J. Physiol.*, **224**, 352–5.
29. BROOKS, S. L., ROTHWELL, N. J., STOCK, M. J., GOODBODY, A. E. and TRAYHURN, P. (1980). *Nature*, **286**, 274–6.
30. ROTHWELL, N. J., STOCK, M. J. and WYLLIE, M. G. (1981). *Biochem. Pharm.*, **30**, 1709–12.
31. ROTHWELL, N. J. and STOCK, M. J. (1981). *Pflugers Archiv.*, **389**, 237–42.
32. COX, J. E. and POWLEY, T. E. (1977). *J. Comp. Physiol. Psychol.*, **91**, 347–58.
33. THURLBY, P. L. and TRAYHURN, P. (1979). *Br. J. Nutr.*, **42**, 377–85.
34. CLEARY, M. P., VASSELLI, J. R. and GREENWOOD, M. R. C. (1980). *Am. J. Physiol.*, **18**, 603–12.
35. KAPLAN, M. L. (1979). *Metabolism*, **28**, 1147–51.
36. TRAYHURN, P., THURLBY, P. L. and JAMES, W. P. T. (1977). *Nature*, **266**, 60–2.
37. TRAYHURN, P. and JAMES, W. P. T. (1978). *Pflugers Archiv.*, **373**, 189–93.
38. TRAYHURN, P. (1979). *Pflugers Archiv.*, **380**, 227–32.
39. THURLBY, P. L. and TRAYHURN, P. (1980). *Pflugers Archiv.*, **385**, 193–201.
40. HIMMS-HAGEN, J. and DESAUTELS, M. (1978). *Biochem. Biophys. Res. Comm.*, **83**, 628–34.
41. GOODBODY, A. E. and TRAYHURN, P. (1981). In: *Proc. Satellite Symposium on Thermal Physiology.* Eds. Szelenyi, Z. and Szekely, M. XXVII Internat. Congr. Physiol. Sci., Pecs., Vol. 32, 515–18.
42. ANDREWS, F. and JACKSON, F. (1978). *Irish J. Med. Sci.*, **147**, 329–30.
43. JUNG, R. T., SHETTY, P. S., JAMES, W. P. T., BARRARD, M. and CALLINGHAM, B. A. (1979). *Nature*, **279**, 322–3.
44. JESSEN, K. (1980). *Acta Anaesth. Scand.*, **24**, 138–43.
45. HEATON, J. M. (1972). *J. Anat.* **112**, 35–9.
46. TANUMA, Y., TAMAMOTO, M., ITOH, T. and YOKOCHI, C. (1975). *Arch. Histol. Japan*, **38**, 43–70.
47. ROTHWELL, N. J. and STOCK, M. J. (1981). *Metabolism*, **30**, 673–8.
48. SMITH, O. L. K. (1978). *Experientia. Suppl.*, **32**, 281–5.

49. ROBINSON, R. P. and PORTE, D. (1973). *Diabetes*, **22**, 1–8.
50. LUDVIGSEN, C., JARRETT, L. and MCDONALD, J. M. (1980). *Endocrinology*, **106**, 786–90.
51. CHINET, A., FRIEDLI, C. and STOCK, M. J. (1980). Unpublished data.
52. DRURY, D. R. (1957). *Arctic. Aeromed. Lab. Tech. Rep.* **46**, 1–10.
53. PERKINS, M. N., ROTHWELL, N. J., STOCK, M. J. and STONE, T. W. (1980). *Nature*, **289**, 401–2.
54. SEYDOUX, J., ROHNER-JEANRENAUD, F., ASSIMACOPOULOS-JEANNET, F., JEANRENAUD, B. and GIRARDIER, L. (1981). *Pflugers Archiv.* **390**, 1–4.
55. LANDSBERG, L., GREFF, L., GUNN, S. and YOUNG, J. B. (1980). *Metabolism*, **29**, 1128–36.
56. GUNDERSEN, H. J. G. and CHRISTENSEN, N. J. (1971). *Diabetes* **26**, 551–7.
57. DANFORTH, E., BURGER, A. G. and WIMPFHEIMER, C. (1978). *Experientia Suppl.*, **32**, 213–8.
58. WIMPFHEIMER, C., SAVILLE, M. E., VOIROL, M. J., DANFORTH, E. and BURGER, A. G. (1979). *Science*, **205**, 1272–3.
59. BURGER, A. G. and SAVILLE, M. E. Personal communication.
60. SAVILLE, M. E. and STOCK, M. J. (1981) *Proc. Nutr. Soc.*, **40**, 59A.
61. ROTHWELL, N. J. and STOCK, M. J. (1981). In: *The Body Weight Regulatory System: Normal and Disturbed Mechanisms*, Eds Cioffi, L., James, W. P. T. and Van Itallie, T. B. Raven Press, NY.
62. NEEL, J. V. (1962). *Am. J. Hum. Gen.*, **14**, 353–7.
63. COLEMAN, D. L. (1979). *Science*, **203**, 663–5.
64. GALE, E. A. M. and TATTERSOLL, R. B. (1978). *Br. Med. J.*, **2**, 1387–9.
65. WISE, P. H. Personal communication.
66. WISE, P. A., EDWARDS, F. M., CRAIG, D. J., EVANS, B., MURCHLAND, J. B., SUTHERLAND, B. and THOMAS, D. W. (1976). *Aust. NZ. J. Med.* **6**, 191–6.
67. BURCH, T. A., BENNETT, P. A. and MILLER, M. (1967). *Diabetes*, **16**, 520–6.
68. MOURATOFF, G. J. and SCOTT, E. M. (1973). *J. Am. Med. Ass.*, **226**, 1345–6.

14

Sweet Taste Receptor Mechanisms

A. FAURION and P. MACLEOD
Laboratoire de Neurobiologie Sensorielle,
92269 Fontenay aux Roses,
France

ABSTRACT

An understanding of taste mechanisms was first expressed in terms of taste quality by chemists and psychologists. Although Henning could already report on the relation between hydroxyl groups and sweet taste in 1915, the chemoreceptor concept was actually introduced only when Renquist in 1918 and Lasareff in 1922 treated mathematically a monomolecular reaction between stimulus and receptor. In 1954 chemoreceptors gained a definitive existence with Beidler's equation which enabled him to conceive from his own electrophysiological data the idea that weak physical forces rather than chemical forces were involved in the taste process.

The problem of the nature of the receptor site still remained and the results of the first single fibre recordings (Pfaffmann, 1941; Beidler, Fishman and Hardiman, 1955; Zotterman, 1949) brought even more complexity to the understanding of the taste process, since single fibres (and single cells) are sensitive to several taste qualities. These results led Pfaffman to suggest in 1959 that the system could work on the principle of an across-fibre pattern. This type of organisation brought the realisation that taste receptors need not be highly specific.

Many decades of chemical modification and the notion of weak binding forces resulted in Shallenberger and Acree's theory (1963, 1967) of sweet taste stimulation based on a double antiparallel hydrogen bonding between the stimulus and the receptor site. Kier, in 1972, added a third feature to the model site involving even weaker forces, i.e. hydrophobic interactions, and complementary steric hindrance is also generally considered to be involved.

However, this model is too restrictive and does not solve the whole question of the nature of the receptor site; we recently reported psychophysical and preliminary electrophysiological data which suggest the existence of several receptor sites (unevenly distributed among individuals) which co-operate in the sweet taste reception mechanism.

Work is presently in progress in this field in order to establish structural features of the site(s), with a major development of quantitative structure activity relationship studies and with the tentative extraction of more or less pure fractions containing the molecule which probably contains the receptor site. A recent idea which makes quantal calculations available for studying the potential intermolecular interaction energy of organic molecules is promising. We are possibly close to understanding the structure and function of the chemoreceptors which enable us to perceive only one sweet taste quality elicited by dozens of different chemical stimuli, some of which have a structure very similar to bitter substances.

INTRODUCTION

Our present knowledge of sweet taste chemoreceptors stems from the intersection of three major disciplines: psychology, chemistry and neurophysiology. Few biophysicists have been interested in taste, and biochemistry was introduced to the field only during the last 20 years.

Even before receptor existence and mechanisms were widely discussed, a large amount of data had already been gathered by psychologists and chemists, the interpretation of which was mostly expressed in terms of taste qualities. Psychologists of the 19th and early 20th century wrote about sweet substances (the stimuli) and sweet quality (the perception). Chemists studied chemical function in relation to sweet taste. Later, electrophysiologists expressed their observations in terms of fibre sensitivity.

It seems that the word chemoreceptor was first used by Beidler in 1953[1] but the concept was already involved in the writings of Renquist[3] and Lasareff[2] who proposed simple equations to account for a monomolecular reaction between the stimulus and, to use Lasareff's own words, 'a sensitive substance contained in the papillae'.

Since then, the word chemoreceptor has been widely used with various meanings; neurophysiologists point to the receptor cell, chemists and biochemists either to the receptor site or to the molecule supporting it.

THE CONTRIBUTION OF THE STRUCTURE–ACTIVITY RELATIONSHIP AND BIOPHYSICAL STUDIES

Psychologists of the 19th century developed psychophysical techniques; 20th century psychophysicists developed, in collaboration with chemists, what are now called structure activity relationship studies. It is surprising that a thorough blending of both domains never occurred: chemists never developed or used sophisticated activity studies but only some hasty tasting techniques, while psychophysicists were satisfied with the use of very few randomly selected sweet substances. As a close collaboration did not exist in the experimental work, a correlation must be looked for from the sparse literature data.

The Qualitative Chemical Approach

As early as 1914 Cohn[4] compiled data on a large number of compounds (6000) to support arguments in favour of the existence of a relationship between constitution and taste. He noticed, on the one hand, that hydroxyl and amino groups are associated with sweet taste, and, on the other hand, that sweet and bitter compounds exhibit several functional groups; he was the first to indicate that what he called 'sapophoric' groups were often found in pairs.

Oertly and Myers in 1919[5] reported that Fischer[6] already knew in 1906 the sweet properties of amino acids and that Henry[7] in 1895 connected the taste of certain bitter compounds with the group $-C(NO_2)CH_2OH$. They themselves tried to supply a simple theory to predict sweet taste from constitution and found 'sweet taste to be dependent on two factors. The glucophores impart to a given compound a potential flavour. If a glucophore is bound to any of the auxoglucs, a sweet compound results.' They defined the glucophore and auxogluc by a long list of functional groups usually found in sweet compounds which did not in itself provide an explanation. They recognised that stereoisomerism is an important factor in the sweetness of amino acids, although they did not acknowledge Piutti[8] who first reported it in 1886.

In 1954 Warfield[9] noticed that sweet taste occurs in compounds having two adjacent features: a group supporting a labile hydrogen and another having a free electron pair.

When Shallenberger[10] studied the relationship between hydrogen bonding in sugars and sweet taste in 1963, quantitative psychophysical studies had already been performed on sweet compounds by Biester et al.,[11] Cameron[12,13] and Schutz and Pilgrim.[14] He found that sweetness is quantitatively related to the inverse of the strength of intramolecular hydrogen bonding.

It was then logical to propose, as Shallenberger and Acree[15] did in 1967, that sweetness should be related to intermolecular hydrogen bonding. Studying the configuration of a number of sugars, they proposed the now generally accepted AH–B system, composed of two antiparallel hydrogen bonds between stimulus and receptor located at about 2·8 Å from each other. This corresponds in glucose, for instance, to two adjacent equatorial OH groups in the staggered conformation, which is in accordance with Warfield's observation. Shallenberger and Acree, however, did not refer to Warfield's work. Extensive investigations of sweet and bitter tasting properties among sugars, sugar derivatives and analogues have been performed by Birch and co-workers[16-20] and they provide strong support for the Shallenberger and Acree model.

A third feature, in the form of a steric hindrance, had soon to be added to the AH–B system in order to account for the sweetness of D-leucine as opposed to the bitterness of L-leucine.[21] The next step was taken by Kier[22] who proposed that it is a hydrophobic group, rather than a steric hindrance, which adequately relates to a wide variety of sweet compounds. As well as being a satisfactory explanation of chirality in sweetness, the hydrophobic third bonding adequately predicts sweetness intensity: the bulkier the hydrophobic moiety, the sweeter the compound, at least within certain limits. Such limits became obvious when quantitative structure–activity relationships were tentatively applied to the sweet taste of large series of chemically related compounds, disclosing unexpected transitions from sweet to bitter taste in amino acids and peptides (Wieser and Belitz,[23] Belitz et al.[24]) and in L-aspartyl peptides (Ariyoshi[25]).

From their results, these authors evolved the idea of a receptor site consisting of the trifunctional AH–B–X system of Kier located at the bottom of a hydrophobic 'pocket'. Closely similar conclusions were almost simultaneously arrived at by Temussi et al.[26] and by van der Heijden et al.[27] (Fig. 1).

The Qualitative Biophysical Approach

Renquist[3] and Lasareff[2] derived simple equations which imply a monomolecular reaction between the stimulus and a substance in the papilla. Lasareff thought that the substance was decomposed by the stimulus which, giving rise to an ionised product, excited the nerves. Papillae sensitive to sweet taste contained a substance which was decomposed by sweet compounds, but this sweet substance could also decompose, with less energy, the substance contained by an acid sensitive papilla, for example.

Beidler[28] in 1954 proposed a theory of taste stimulation on the basis of electrophysiological integrated recordings of the taste nerves.[29] He assumed that the neural response is proportional to the number of stimulus molecules bound to acceptor sites at equilibrium. As in the earlier papers of Renquist and Lasareff the adsorption is described as a monomolecular reaction obeying the mass action law. Expressing his equilibrium equation in the form of C/R (concentration/response) as a function of C (concentration), Beidler found a relationship $(C/R = C/R_m + 1/KR_m$, where R_m is the maximum response) which is similar to a Langmuir's adsorption isotherm. Plotting his own results in these terms $(C/R = f(C))$ he deduced that the association constant (K) is very low, and calculated free energy variations $(\Delta F = -RT \log K)$ of the order of 1 to 2 kcal/mol for sodium salts. This implied that low physical rather than chemical binding energies are involved between stimulus and receptor.

Dzendolet,[30] pointing to the fact that Beidler's data do not fit the model equation at low concentrations, proposed a small modification taking into account the number of free molecules rather than the whole number of stimulus molecules. His equation fits Beidler's data, even at low concentrations of the stimulus.

Heck and Erickson[31] developed a complementary theory of taste stimulation taking into account the rate of stimulus adsorption in addition to the proportion of occupied sites, in order to describe the transient part of the neural responses, as well as the steady states.

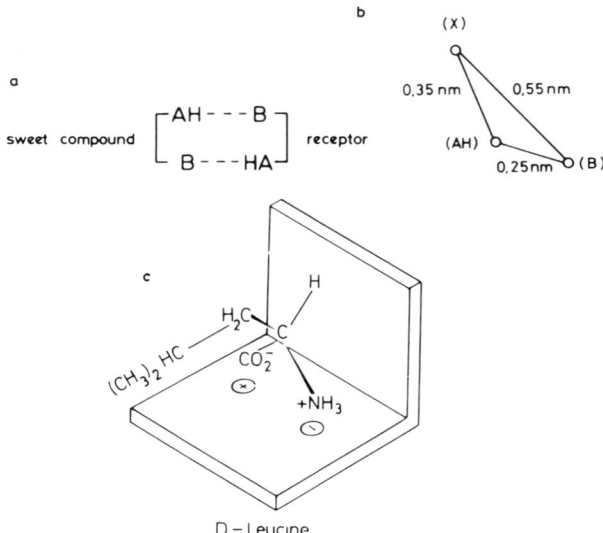

FIG. 1. Evolution of the model of the sweet receptor site. (a) Bifunctional site of Shallenberger;[15] (b) the trifunctional site of Kier;[22] (c) steric hindrance added to the bifunctional site can account for discrimination between D-and L-amino acids.[21]

The Direct Biochemical Approach

If chiral sweet receptor sites actually exist in the taste cell membrane, then it should be possible to extract some, presumably proteinaceous, material exhibiting specific affinity for sweet molecules.

Dastoli and Price[32] apparently succeeded in extracting from bovine tongue a 'sweet-sensitive protein' which exhibited the expected different absorption spectra in the presence of sugars and saccharin. However, it soon became clear that this protein was not specifically related to the taste process.[33]

Hiji and Sato[34] extracted protein fractions which were not present in denervated tongues when taste buds had degenerated. Kadowaki et al.[35] analysed fractions obtained from fungiform, circumvallate papillae and tongue epithelium of dog, monkey and man by electrophoresis in SDS acrylamide gels. They found some protein bands specifically belonging either to fungiform or to circumvallate papillae, with some variation among individuals. This is indicative of the existence of different proteins in both kinds of papillae but their functions have yet to be elucidated.

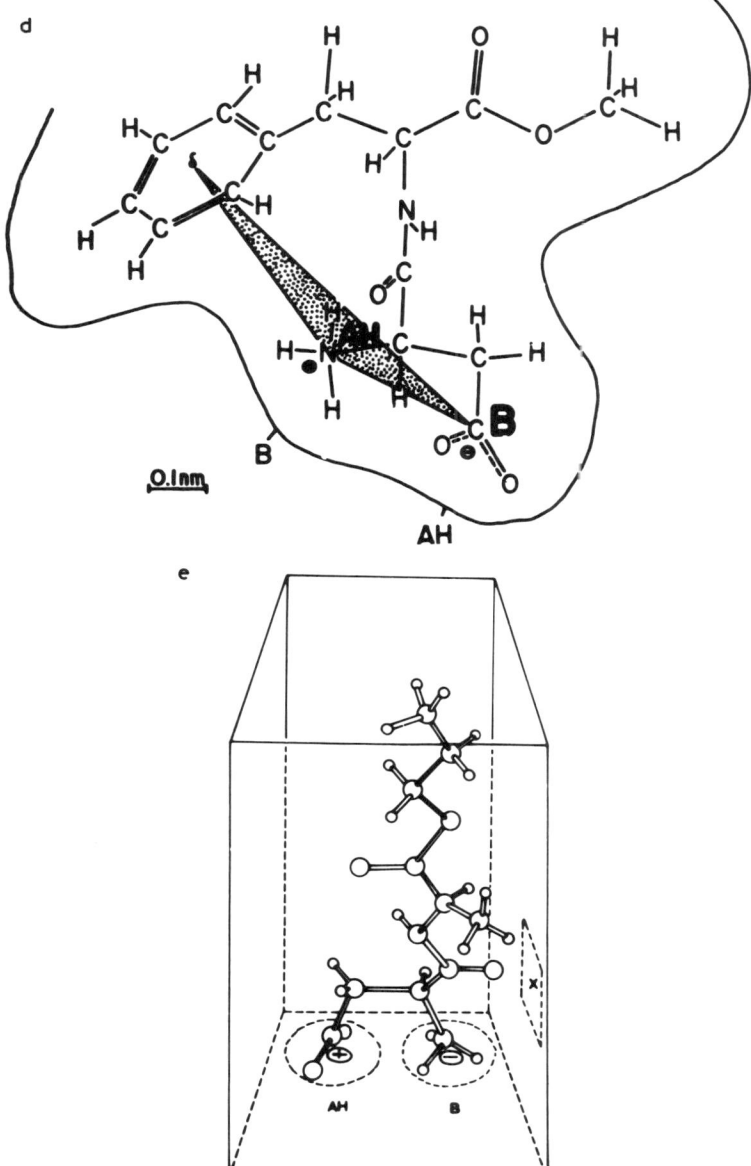

FIG. 1.—*contd.* Pocket versions of the trifunctional site: (d) var. der Heijden *et al.*,[27] aspartame; (e) Ariyoshi,[25] L-asp-D-ala-OPr[a].

Cagan[36] and Lo,[37] in well-controlled experiments, showed that homogenates of bovine circumvallate and fungiform papillae bind more ^{14}C-labelled sucrose, glucose and fructose (but not lactose) than do homogenates of filiform papillae. Cagan and Morris[38] obtained similar results with ^{3}H-labelled methyl-monellin on bovine and human circumvallate papillae homogenates. They further obtained a displacement of 20–40% of the bound monellin through competitive interaction with sucrose, lactose, saccharin, cyclamate, neohesperidin dihydrochalcone and aspartame but not with stevioside. It may thus be concluded that taste buds contain some specific material which binds sweet-tasting ligands.

Indirect supporting evidence is provided by Hiji,[39] who selectively suppressed the taste response to sugars in rat and man by applying pronase E to the tongue, and by Sato et al.[40] who showed that Japanese macaques exhibit parallel variations in their individual preferences, electrophysiological responses and *in vitro* binding towards a variety of sweeteners.

Nevertheless up to the present time the scarcity of binding material and the weakness of the binding energies have prevented anyone from convincingly isolating a pure sweet receptor substance. It can be anticipated, however, that the availability of clonal antibodies will afford a better opportunity of completing the direct characterisation of taste receptor sites, possibly *in situ* rather than *in vitro*. In the former case, the function would not be disrupted and could be tested with neurophysiological or psychophysical techniques.

THE CONTRIBUTION OF NEUROPHYSIOLOGICAL AND PSYCHOLOGICAL STUDIES: FUZZY PATTERNS

When electrophysiological techniques came within the reach of taste researchers, they confused the issue still further as they revealed a great complexity in the mechanism of neural coding of taste. First recordings of unitary fibres of chorda tympani had been performed by Zotterman in 1935,[41] Pfaffmann in 1939[42,43] and Beidler et al. in 1955 and subsequently.[44,45] These recordings showed that a single fibre was sensitive to several taste qualities (Fig. 2) and researchers had then to forsake the narrow view of a specific organ devoted to each of the four qualities; salty, acid,

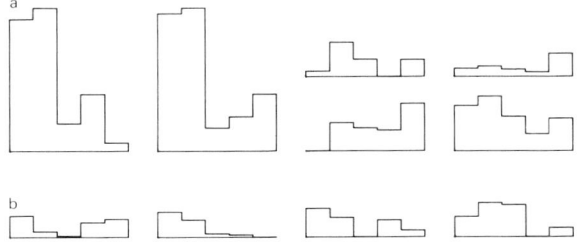

FIG. 2. Differential sensitivity of chorda tympani single units Number of spikes elicited by taste stimulation. From left to right, in each histogram, HCl, KCl, NaCl, quinine and sucrose. (a) 6 units in the rabbit; (b) 4 units in the cat. Redrawn from Pfaffmann.[83]

sweet and bitter. Furthermore, Kimura and Beidler[46, 47] showed by intracellular recordings that the sensory cell itself is sensitive to more than one quality, suggesting that chemoreceptors are more or less randomly distributed on the membranes of sensory cells.

The Objective Electrophysiological Approach

Pfaffmann, in 1939, had recorded the response of unitary fibres in a cat in order to classify them and at best could only distinguish three main types, namely: acid, NaCl–acid and quinine–acid responding fibres. He then concluded:

'These fibre categories are based, of course, upon the four types of stimuli used for testing. There is evidence that certain other substances may also stimulate more than one fibre, so that if a wide variety of agents were used, each fibre might be found to have a chemical "spectrum" which overlapped those of other fibres.'

In 1959 his conclusion was as follows:

'It is clear, from recent electrophysiological evidence, that the taste receptors do not always fall into four basic receptor types corresponding to the basic taste qualities. The individual sensory cells are differentially sensitive to chemicals, probably because of differences at sites on the cell membrane. The chemical specificity of the taste cell can best be described as a cluster of sensitivities which varies among different receptor cells. Any one cell is reactive to a varying degree to a number of different chemical stimuli, many of which fall in two or more of the four classical basic taste categories.'[48] (See also Fig. 3.[49])

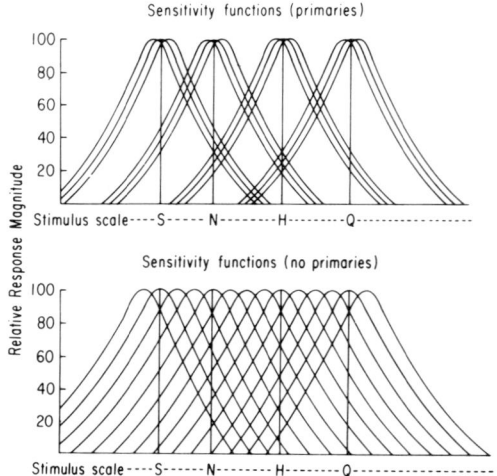

FIG. 3. Taste continuum with or without clusters of sensitivities. Each curve represents the amplitude variation of the response of a given taste fibre to the series of stimuli arranged in a putative continuum along the abscissae. Two hypotheses are sketched: four overlapping continua arranged around the four primaries, or, a single continuum covering the whole range of taste without primaries. From Pfaffmann.[49]

Working out the responses of 79 single taste fibres to 12 stimuli in the hamster, Frank[50] found that sensitivities to sucrose, fructose and saccharin were strongly associated with each other and almost independent of other non-sweet sensitivities. She accordingly emphasised the concept of fibres responding best to each of the four basic tastes, despite the fact that response profiles to non-sweet stimuli are substantially correlated to each other.

Erickson developed the expression of an across-fibre pattern to point out that the stimulus is probably signalled by 'a pattern of activity in a number of fibre groups rather than by the activity in any single fibre group alone'. A large amount of electrophysiological data supports this concept.[51-56]

The difference which remains between Pfaffmann and Erickson at the present time is that Pfaffmann stresses the existence of clusters of sensitivities,[57] while Erickson fails to find neural response profile types and hence argues against the 'concept of primaries or types at the neural level' at least in the salt–acid range.[58]

An important contribution of neurophysiology is the evidence, provided by Hellekant et al.,[59] that the sweet proteins monellin, thaumatin and miraculin are intensely sweet only to some primates and man but not to rodents, dog or pig. This clearly raises the question of different acceptors contributing to one and the same sweet taste response among different mammals and possibly coexisting in the same species.

Another puzzling finding is that chorda tympani fibres respond to warm stimulation in hamster whereas fibres responsive to acid and/or quinine respond to cold as well.[60]

The Subjective Psychophysical Approach

The idea that some tastes are opposed, derived or intermediate was proposed by Aristotle:[61]

'one can distinguish among tastes as among colours, on the one hand, simple species which are opposite, that is, sweet and bitter, and on the other hand, species which are derived either from the first one such as succulent, or from the second one such as salty and finally, species which are intermediate between these last tastes: sour, harsh, astringent and acid.'

For many centuries tactile and taste sensations were thought of as being indistinguishable. Linnaeus[62] who classified the gustatory substances through their medical properties, identified 10 qualities: wet, dry, acid, bitter, fat, astringent, sweet, sharp, mucous, salty.

In 1824, Chevreul[63] accurately distinguished between olfactory, gustatory and tactile sensitivity. The four qualities were recognised in the 19th century; Henning[64] quotes Fick[65] as probably being the first to support this classification.

In 1914, Cohn asserted that the four qualities were isolated modalities. Henning presented opposing arguments in 1916 in favour of a continuum (kontinuierliche Reihe), which he illustrated using the surface of a tetrahedron as a model. Quoting in the main Kiesow's experimental work and discussion, he placed ammonium chloride and potassium aluminate between salty and acid, potassium bromide and iodide between salty and bitter, alkaline between salty and sweet and so on. He already knew that the hydroxyl group was involved in sweet taste and explained alkaline taste in the following way:

'lyes have the same anion as salt and their second moiety, an hydroxyl group, is the sweet group. So papillae corresponding to salt taste and to sweet taste react under the stimulation by lyes.'

In their pioneering work, Yoshida and Saito[66,67] investigated taste quality perception by means of multidimensional scaling of pair similarity judgments on 17 amino acids at 32 times their threshold concentration in 15 subjects. They found a nearly Euclidean, three-dimensional space in which each amino acid occupied a specific position. It was easy to divide this cluster into three subsets respectively corresponding to bitter, sweet and sour dominant ratings in a parallel profile study. It was noticed that judgments of individual subjects differed significantly from each other, albeit self-consistently. They nevertheless complied with contemporary usage in analysing only the grouped data (Fig. 4a). Moreover they superimposed upon their spatial representation a Henning tetrahedron that is certainly not the best fitting simple geometrical figure.

The concept of a continuum has recently been revived by Erickson[68,69] who, following Pfaffmann[48], supports the view of a continuum of sensitivity based upon electrophysiological as well as psychological data. Studies based on subjects' judgments of similarity among stimuli show that 'the spatial arrangement of taste stimuli derived from multidimensional scaling procedures cannot be totally contained within the salty, sweet, sour and bitter domain represented by a tetrahedron' (Fig. 4b). Schiffman, Erickson and co-workers have observed that:[70-74]

'sweet, sour, salty and bitter do not even span the whole range of taste.'

'Even if àll taste stimuli could be described in terms of sweet, sour, salty and bitter, this would not demonstrate that the system is analytic with four unique sensations or that four is the appropriate number'

As a matter of fact, Erickson[73] has shown that a typical sensory continuum such as vision could be described in terms of four or five standards. Many authors have been misled by using only four taste stimuli; Pfaffmann had already alluded to this reductive procedure as early as 1941.

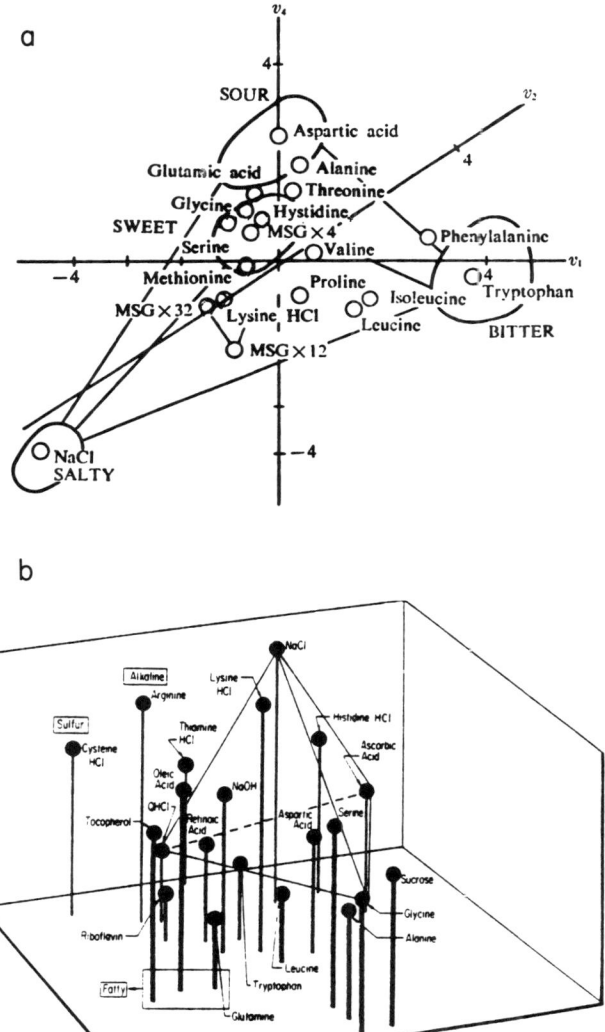

FIG. 4. Multidimensional scaling of taste similarities of a variety of stimuli tentatively covering the whole taste range. (a) From Yoshida and Saito;[66] (b) from Schiffman and Dackis.[71] The three-dimensional taste space does not fit Henning's tetrahedron.

When a continuum is considered, one can ask whether it is analytic or synthetic, that is, if the qualities are mixed, is it still possible to distinguish each of them in the mixture, or do they merge into a new different quality? In his model, Henning emphasises the fact that single tastes exist along the edges and surface of his tetrahedron. Concerning Kiesow's problem of whether alkaline taste is a composition of the four qualities or a fifth one, he proposes a third possibility, namely that 'alkaline is a simple transitional taste which shows similarity to salty as well as to sweet taste'. Erickson and Covey[74] compared vision, audition and gustation in this respect. Audition appears fully analytic, vision is synthetic and gustation is between the two: a so-called pure taste ('primary stimulus') is judged unitary only 50% to 85% of the time and a mixture of sucrose and NaCl is judged unitary 39% of the time.

INTERINDIVIDUAL DIFFERENCES: A STUMBLING BLOCK OR AN OPENING?

Interindividual Differences of Subjects' Sensitivity for Various Sweeteners

Experiments have been conducted since 1974 in our laboratory and these have shown interindividual differences in sweet taste sensitivity on the basis of quantitative psychophysical measurements.[75, 76]

A threshold evaluation for seven compounds among 98 subjects showed that individual threshold profiles were all different. In other words, knowing the threshold values of subject A for two sweet products and the threshold value of subject B for one of them, will not permit deducing subject B's threshold for the second compound.

Interindividual differences were similarly found at suprathreshold level. Intensity measurements were plotted against concentration for nine subjects and 12 sweeteners. The experimental procedure consisted in matching the intensity of seven concentrations of each sweetener against a sucrose reference scale. Linear regressions of intensity on concentration were calculated and the slopes appear to depend both on sweetener and subject. Fifty per cent of the slopes were significantly different from one another and their ranking varied independently from subject to subject.

In both studies, correlation coefficients have been calculated across stimuli. At threshold level, they are all below 0·37, which indicates an impressive independence between subjects' sensitivities to the seven sweeteners. At suprathreshold level, correlations range from $-0·8$ to $+0·8$. The same pairs are best correlated at both levels except for saccharin.

Multidimensional analysis (factor analysis of Benzécri et al.[77]) has been performed on both sets of data. Our results presented in Fig. 5 consistently show that at both levels thaumatin, sucrose and D-leucine do not covary; chloroform, tested at suprathreshold level only, is another non-covarying stimulus.

As far as subjects are concerned, it can be seen, especially at threshold level, that they are displayed throughout a continuum. Our results are in accordance with those of Schiffman[78] obtained on a basis of judgments of quality similarities. Slight differences can be accounted for by side-tastes or off-flavours involved in her measurements, which are eliminated in ours.

Differential Inhibitory Effect of Pronase E on Sweet Sensitivity in Man

The reduction of sweet-taste intensity by pronase E has been measured for 20 sweeteners on five subjects, with an intensity matching procedure.[79] As can be seen in Fig. 6, some subjects are more sensitive than others to the same dose of inhibitor; but more interesting is the fact that individual profiles of inhibition do not covary. For example, saccharin is inhibited in subject 5 but not in subject 2, while monellin is inhibited in subject 2 but not in subject 5.

These data, subjected to the same factor analysis, display the stimuli in a three-dimensional space which is coherent, on the whole, with the gustatory space obtained in the preceding two experiments: D-leucine, thaumatin and sucrose do not covary together. Supplementary information is provided such as the confirmation that saccharin and monellin do not covary either with the other sweeteners.

The same experiment, using an extract of the leaves of *Gymnema sylvestre* instead of pronase E, led to a similar conclusion. A noticeable feature is that chloroform is uniformly and weakly inhibited for every subject as opposed to the other stimuli.

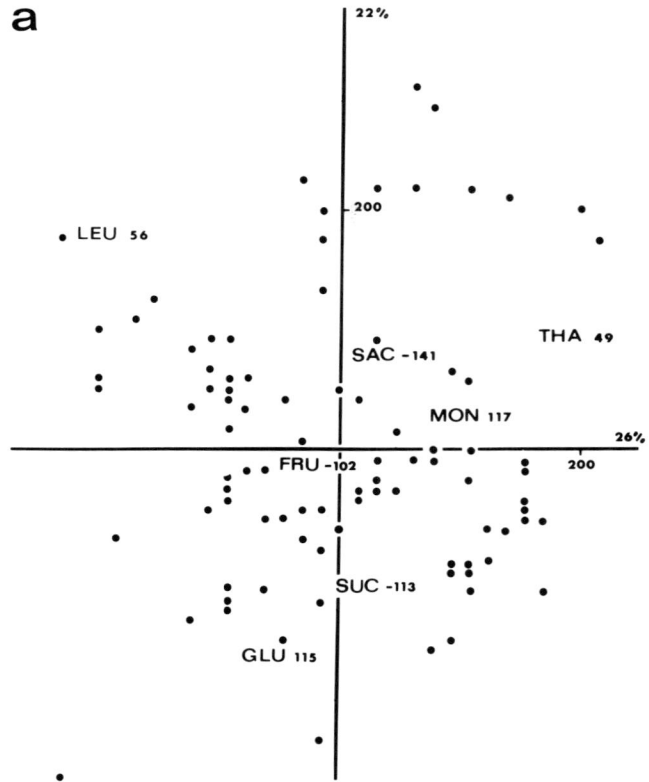

FIG. 5. Factor analysis of individual quantitative responses to sweet stimuli. This analysis associates inter-stimulus vector distances to the amount of covariance between subjects' responses on the basis of χ^2 calculation. Similarly, and in the same space, subjects are displayed according to the covariance of their responses to the stimuli. The smaller the covariance, the larger the distance.[77] Subjects are represented by points, stimuli by alphabetic symbols which are accompanied by the third axis coordinates: (ASP, aspartame; CHL, chloroform; CYC, Na-cyclamate; DUL, dulcin; FRU, D-fructose; GLU, D-glucose; LEU, D-leucine; MON, monellin; SAC, saccharin; STV, stevioside; SUC, sucrose; THA, thaumatin). (a) Individual recognition thresholds; 80% of the information is accounted for by four factors. Measurements were not replicated.

Electrophysiological Recordings of Single Units

A new technique has been developed to record neural activity of single units through the pore of the papilla with glass micropipettes, while a neighbouring papilla is stimulated.

Chorda tympani fibres have been known to be branched, since

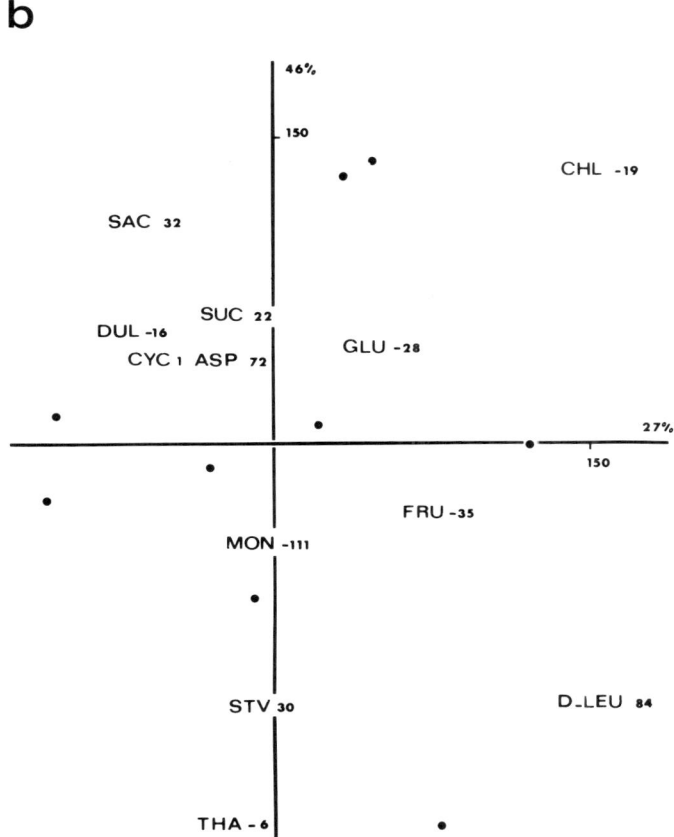

FIG. 5.—contd. (b) Individual slopes of the linear regression functions of intensity on concentration; 86% of the information is accounted for by three factors. Slope values were computed from 100-150 elementary responses per subject and sweetener.

the studies reported by Miller,[80] and are known to innervate one to nine taste buds. Oakley[81] studied receptor fields of single chorda tympani fibres and found that connected papillae lay most commonly 2 mm apart from each other in the cat, and sometimes as far apart as 18 mm. Our recording technique hopefully takes advantage of this anatomical feature for reducing the number of chemoreceptors contributing to the recorded activity.

Besides their already known differential sensitivity to various

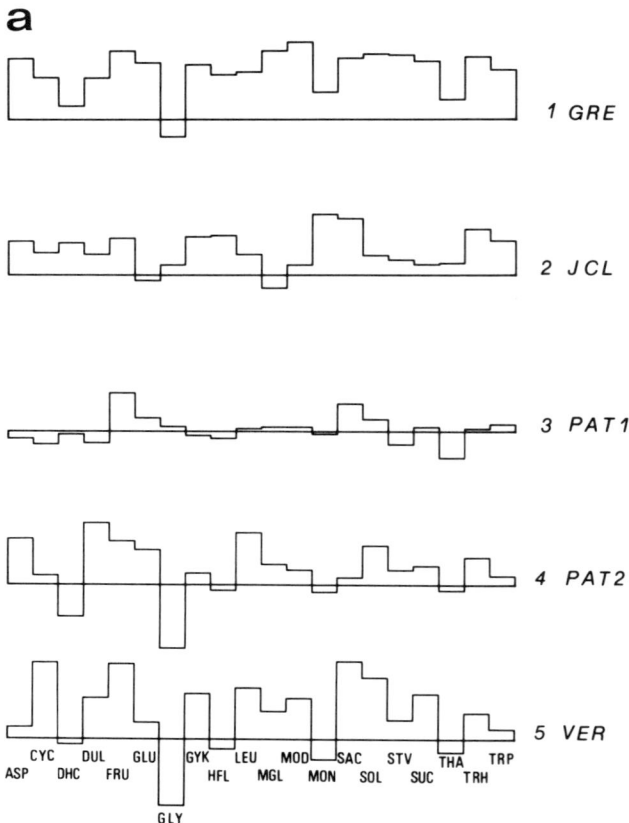

FIG. 6. Differential inhibitory effect of pronase E on the sensitivity of five subjects to 20 sweeteners. (a) Individual profiles of sweetness magnitude used to estimate differences before and after application of pronase E on the tongue. Sweetness reduction after pronase E is plotted upwards. All sweeteners are inhibited except glycyrrhizic acid which is significantly reinforced in three subjects. (25 replications).

taste qualities, taste fibres are shown to possess response spectra to sweet stimuli, which vary independently from unit to unit (Fig. 7).[79]

Structural Studies

The above results seem to indicate that sweet taste is actually a multiple sensitivity and that any sweet stimulus is likely to be adsorbed by a set of various different 'sweet' receptor sites with a

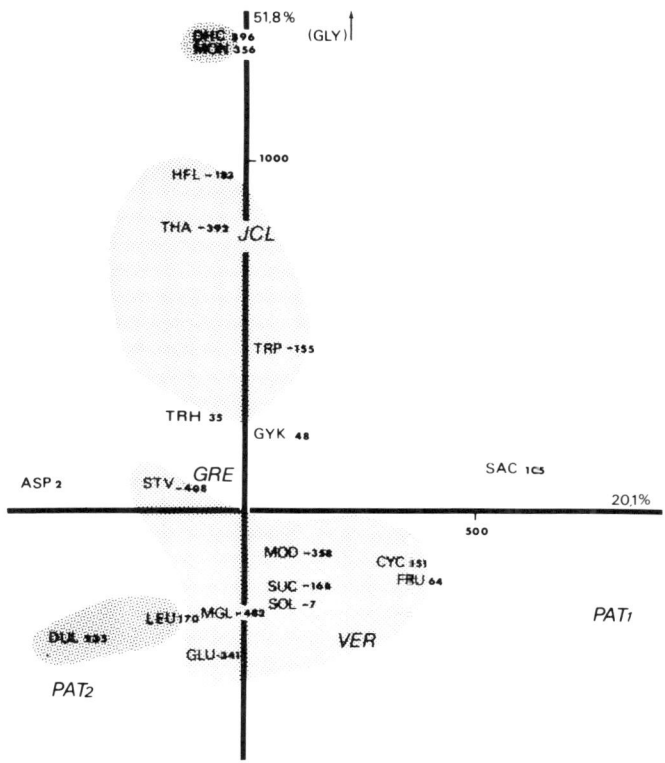

FIG. 6.—*contd.* (b) Factor analysis of data presented in (a); 88% of the information is accounted for by three factors. Stippled areas outline groups of stimuli which are strongly pair-correlated (Pearson's $r > 0.9$). ASP, aspartame; CYC, Na-cyclamate; DHC, neohesperidin dihydrochalcone; DUL, dulcin; FRU, D-fructose; GLU, D-glucose; GLY, glycyrrhizic acid; GYK, glykergenic acid; HFL, 4β, 10α-dimethyl-1,2,3,4,5,10-hexahydrofluorene-4α, 6-dicarboxylic acid; LEU, L-leucine; MGL, methyl α-D-glucopyranoside; MOD, acetosulfam; MON, monellin; SAC, saccharin; SOL, D-sorbitol; STV, stevioside; SUC, sucrose; THA, thaumatin; TRH, α, α-trehalose; TRP, D-tryptophan.

rather low specificity. The stimuli may interfere preferentially with some given sites and individual differences among subjects may result from genetic differences in the proportion of these sites. Whether a large number of sharply tuned sites or a smaller number

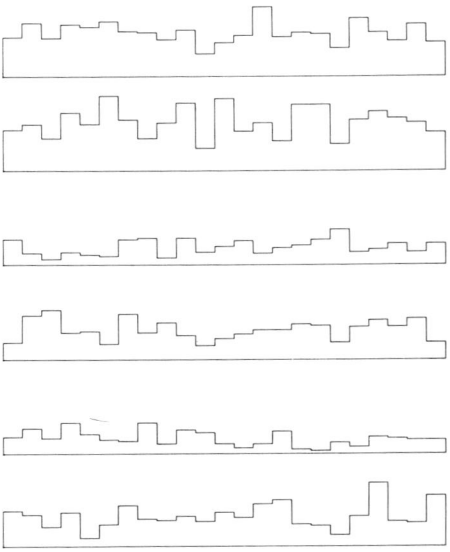

FIG. 7. Response spectra of six single units recorded in hamster fungiform papillae. Stimuli are the same as in Fig. 6, with three more added: BZX, 2-(3-hydroxy-4-methoxyphenyl)-1,3-benzodioxan; CHL, chloroform; MAN, 1-methyl-2-amino-4-nitrobenzene; all ranged in alphabetical order. Responses are expressed in terms of supplementary spikes during 4 s after stimulation: $R = O - C$, where O is the observed number of spikes and C a mean calculated during 40 s preceding the stimulation. Plotted data are a final mean of two to five responses to each of the 23 sweeteners, for each unit.

of broadly tuned sites can be expected, cannot be decided on the basis of these results.

Hence, it appears that the models of sweet receptor sites which successively appeared in the literature are too restricted in the sense that they reduce sweet taste to a monadic sensitivity: their authors have looked for common properties of a variety of sweet compounds and proposed a unique acceptor. Starting with the models more recently proposed and assuming that they are valid as far as the kind of bonding and energies are concerned, it is suggested that several geometrically different receptor sites be sought in order to accommodate the great variety of sweet stimuli.

In biological activity studies dealing with a large enough number of sweeteners, we find independent groups of stimuli which are confirmed by paired correlation coefficients. It is therefore approp-

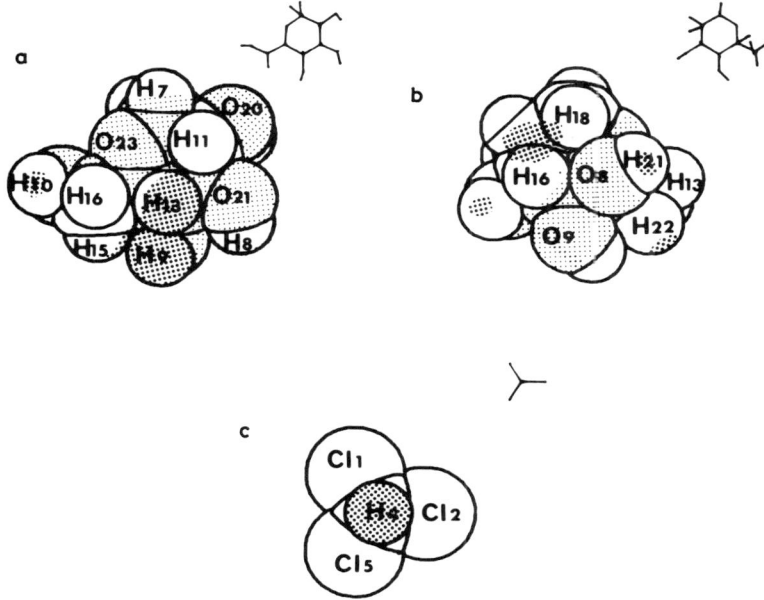

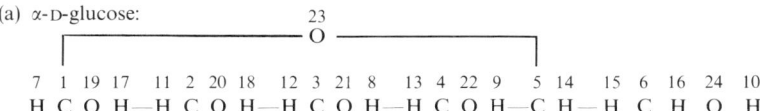

FIG. 8. Orthogonal projections of the van der Waals contours of 6 sweet tasting molecules and a non sweet structural analogue. Light stippled areas, electronegative zones; heavy stippled, electropositive zones; unshaded areas, hydrophobic zones.

(a) α-D-glucose:

```
                              23
          ┌─────────────────── O ───────────────────────────────┐
    7 1 19 17   11 2 20 18   12 3 21 8   13 4 22 9   5 14   15 6 16 24 10
    H C O H   — H C O H — H C O H — H C O H — C H — H C  H  O  H
```

(b) β-D-fructopyranose:

```
                                   12
          ┌─────────────────────── O ──────────────────────────┐
    13 1 14 7 20   2 8 21   15 3 9 22   16 4 10 23   17 5 11 24   18 6 19
    H C H O H — C O H — H C O H — H C O H — H C O H — H C H
```

(c) Chloroform

riate to try to characterise sweet molecules in terms of low energy potentials of interaction, to look for groups on the basis of common structural features (which can be regarded as indications

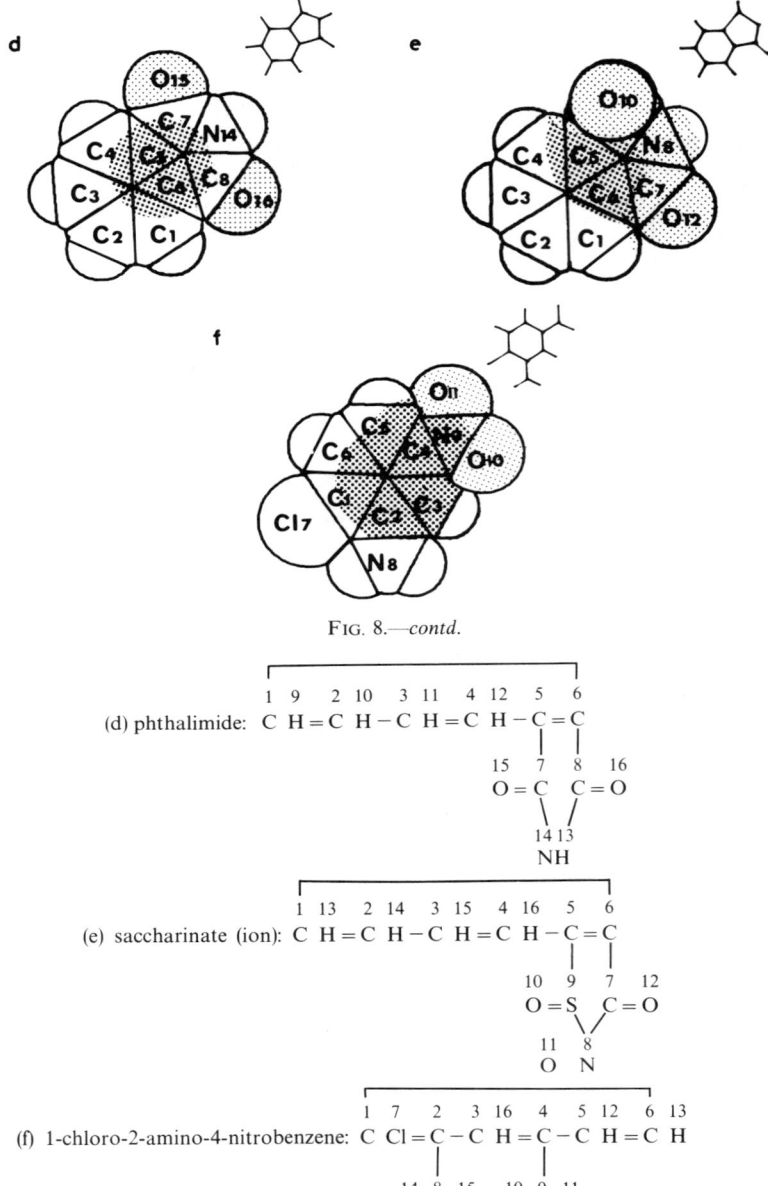

FIG. 8.—contd.

$$
\text{(d) phthalimide:} \quad \overset{1\ \ 9}{\text{C H}} = \overset{2\ \ 10}{\text{C H}} - \overset{3\ \ 11}{\text{C H}} = \overset{4\ \ 12}{\text{C H}} - \overset{5}{\underset{|}{\text{C}}} = \overset{6}{\underset{|}{\text{C}}}
$$

$$
\overset{15}{\text{O}} = \overset{7}{\underset{\diagdown}{\text{C}}} \quad \overset{8}{\underset{\diagup}{\text{C}}} = \overset{16}{\text{O}}
$$

$$
\underset{14\ 13}{\text{NH}}
$$

$$
\text{(e) saccharinate (ion):} \quad \overset{1\ \ 13}{\text{C H}} = \overset{2\ \ 14}{\text{C H}} - \overset{3\ \ 15}{\text{C H}} = \overset{4\ \ 16}{\text{C H}} - \overset{5}{\underset{|}{\text{C}}} = \overset{6}{\underset{|}{\text{C}}}
$$

$$
\overset{10\ \ 9}{\text{O} = \text{S}} \quad \overset{7\ \ 12}{\text{C} = \text{O}}
$$

$$
\underset{11\ \ 8}{\text{O} \ \ \text{N}}
$$

$$
\text{(f) 1-chloro-2-amino-4-nitrobenzene:} \quad \overset{1\ \ 7}{\text{C Cl}} = \overset{2}{\underset{|}{\text{C}}} - \overset{3\ \ 16}{\text{C H}} = \overset{4}{\underset{|}{\text{C}}} - \overset{5\ \ 12}{\text{C H}} = \overset{6\ \ 13}{\text{C H}}
$$

$$
\underset{14\ \ 8\ \ 15}{\text{H N H}} \quad \underset{10\ \ 9\ \ 11}{\text{O N O}}
$$

of common receptor sites) and to cross-check them with the groups obtained from psychophysical or electrophysiological data.

Studies are being developed in our laboratory for this purpose: in order to describe sweet molecules and their non-sweet analogues in terms of low energy binding potentialities, the energy of interaction of the molecule with a probe which can be H^+, H^- or H_2O, is calculated on the basis of simplified formulae.[82] The interaction

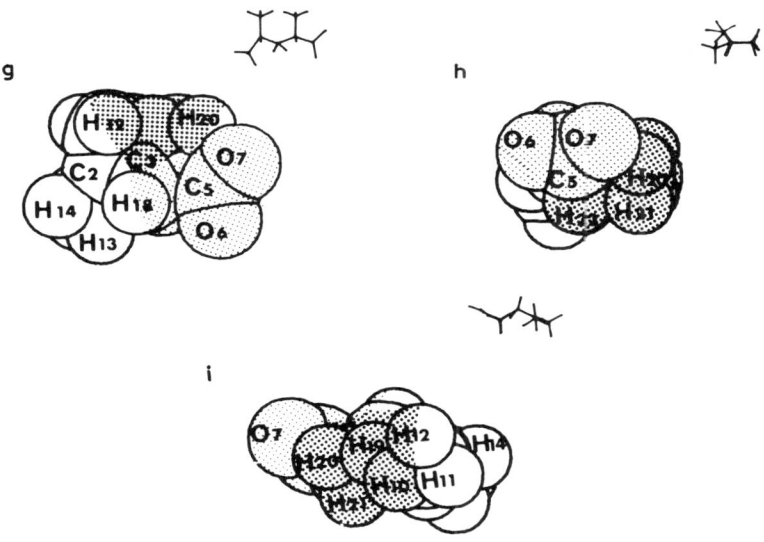

FIG. 8.—contd. (g) (h) (i) Three orthogonal views of D-leucine.

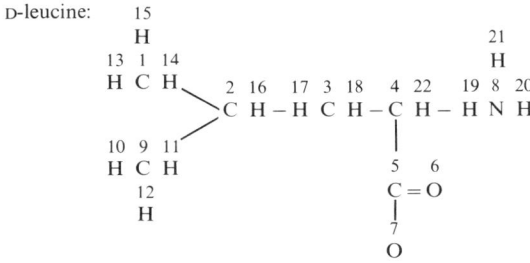

energy all around the molecule is calculated and the contact points obtained after minimisation of the energy are projected onto the surface of the molecule. Projection of contacts with H^+ depicts electronegative areas, with H^-, electropositive areas; hydrophobic areas are depicted by energies of interaction with water molecules lower than 3 kcal/mol. Figure 8 shows the possibilities of such an approach. In our results flat molecules such as saccharin are so depicted that they are fully compatible with Kier's site. Other molecules such as D-leucine are better accommodated by 'pocket' sites as their three-dimensional structure must be accounted for: the negative, positive and hydrophobic areas do not lie on the same face. As far as chloroform is concerned, it appears in this calculation to be lacking electronegativity. It is not the only peculiarity of this molecule since its behaviour in psychophysical tests is different from that of other molecules.

These are only tentative results but our opinion is that this approach is promising in that it allows a topographical characterisation of the level of energies which are actually involved in the chemoreception process.

REFERENCES

1. BEIDLER, L. M. (1953). *J. Neurophysiol.*, **16**, 595–607.
2. LASAREFF, P. (1922). *Archiv fur die Gesamte Physiologie des Menschen und der Tiere*, **194**, 293–7.
3. RENQUIST, Y. (1919). *Skand. Arch. für Physiol.*, **38**, 97–201.
4. COHN, G. (1914). *Pharm. Zentral.*, **55**, 735–47.
5. OERTLY, E. and MYERS, R. G. (1919). *J. Am. Chem. Soc.*, **41**, 855–67.
6. FISCHER, E. (1906). *Ber. d. Chem. Ges.*, **39**, 2320–8.
7. HENRY, C. (1894). *C. R. Soc. Biol.*, **1**, 682.
8. PIUTTI, A. (1886). *C. R. Acad. Sc. Paris.*, **103**, 134–7.
9. WARFIELD, R. B. (1954). In: *Abstracts of the 126th Meeting of the Am. Chem. Soc. Div. Agr. Food Chem.* New York p. 15a. (abstract).
10. SHALLENBERGER, R. S. (1963). *J. Food Sci.*, **28**, 584–9.
11. BIESTER, A., WOOD, M. W. and WAHLIN C. S. (1925). *Am. J. Physiol.*, **73**, 387–96.
12. CAMERON, A. T. (1943). *The Taste Sense and the Relative Sweetness of Sugars and Other Sweet Substances*, Sugar Research Foundation Scientific Report Series, Number 9, New York.
13. CAMERON, A. T. (1943). *Trans. R. Soc. Can.*, **37**, 11–27.
14. SCHUTZ, H. G. and PILGRIM, F. J. (1957). *Food Res.*, **22**, 206–13.
15. SHALLENBERGER, R. S. and ACREE, T. S. (1967). *Nature*, **216**, 480–2.

16. BIRCH, G. G. and LINDLEY, M. G. (1973). *J. Food Sci.*, **38**, 665–7.
17. LINDLEY, M. G. and BIRCH, G. G. (1975). *J. Sci. Fd Agrc.*, **26**, 117–24.
18. LEE, C. K. and BIRCH, G. G. (1975). *J. Food Sci.*, **40**, 784–7.
19. BIRCH, G. G., and LEE, C. K. (1976). *J. Food Sci.*, **41**, 1403–7.
20. BIRCH, G. G. (1976). *CRC Critical Reviews in Food Science and Nutrition*, **8**, 57–95.
21. SHALLENBERGER, R. S., ACREE, T. S. and LEE, C. K. (1969). *Nature*, **21**, 555–6.
22. KIER, L. B. (1972). *J. Pharm. Sci.*, **61**, 1394–7.
23. WIESER, H. and BELITZ, H. D. (1975). *Z. Lebensm. -Untersuch. -Forsch.* **159**, 65–72.
24. BELITZ, H. -D., CHEN, W., JUGEL, H., TRELEANO, R., WIESER, H., GASTEIGER, J. and MARSILI, M. (1979). In: *Food Taste Chemistry*, Ed. Boudreau, J. C. American Chemical Soc. Washington, 93–131.
25. ARIYOSHI, Y. (1979). In: *Food Taste Chemistry*, Ed. Boudreau, J. C., American Chemical Society, Washington, 133–48.
26. TEMUSSI, P. A., LELJ, F. and TANCREDI, T. (1978). *J. Med. Chem.*, **21**(11), 1154–8.
27. VAN DER HEIJDEN, A., BRUSSEL, L. B. P. and PEER, H. G. (1979). *Chem. Senses Flavor*, **4**(2), 141–52.
28. BEIDLER, L. M. (1954). *J. Gen. Physiol.*, **38**, 133–9.
29. BEIDLER, L. M. (1953). *J. Neurophysiol.*, **16**, 595–607.
30. DZENDOLET, E. (1967). *Perception and Psychophysics*, **2**, 519–20.
31. HECK, G. L. and ERICKSON, R. P. (1973). *Behav. Biol.*, **8**, 687–712.
32. DASTOLI, F. R. and PRICE, S. (1966). *Science*, **154**, 905–7.
33. KOYAMA, N. and KURIHARA, K. (1971). *J. Gen. Physiol.*, **57**, 297–302.
34. HIJI, Y. and SATO, M. (1973). *Nature New Biology*, **244**, 91–3.
35. KADOWAKI, H., SASAKI, H., YAMADA, H. and HIJI, Y. (1979). In: *Proceedings of the 13th Japanese Symposium on Taste and Smell*, Ed. Funakoshi, M. Department of Oral Physiology. Gifu College of Dentistry, Gifu, 95–8.
36. CAGAN, R. P. (1971). *Biochim. et Biophys. Acta*, **252**, 199–206.
37. LO, C. H. (1973). *Biochim. et Biophys. Acta*, **291**, 650–61.
38. CAGAN, R. P. and MORRIS, R. W. (1979). *Proc. Nat'. Acad. Sci. USA*, **76**, (4), 1692–6.
39. HIJI, Y. (1975). *Nature*, **256**, 427–9.
40. SATO, M., HIJI, Y., ITO, H., IMOTO, T. and SAKU, C. (1977). In: *Food Intake and Chemical Senses*, Eds Katsuki, Y., Sato, M., Takagi, S. F. and Oomura, Y., Japan Scientific Societies Press, 187–99.
41. ZOTTERMAN, Y. (1935). *Skand. Arch., für Physiol.*, **72**, 73–7.
42. PFAFFMANN, C. (1939). *J. Physiol.*, **96**, 41P–42P, (Abstract).
43. PFAFFMANN, C. (1941). *J. Cellular and Comparative Physiology*, **17**, 243–58.
44. BEIDLER, L. M., FISHMAN, I. Y. and HARDIMAN, C. W. (1955). *Am. J. Physiol.*, **181**, 235–9.
45. FISHMAN, I. Y. (1957). *J. Cell. Comp. Physiol.*, **49**, 319–34.
46. KIMURA, K. and BEIDLER, L. M. (1956). *Am. J. Physiol.*, **187**, 610–11.

47. KIMURA, K. and BEIDLER, L. M. (1961). *J. Cellular and Comparative Physiology*, **58**, 131–9.
48. PFAFFMANN, C. (1959). In: *Handbook of Physiology—Section I, Neurophysiology*. Eds Field, J. and Magoun, H. W. American Physiological Society, Washington, 507–33.
49. PFAFFMANN, C. (1974). *Chem. Senses Flavor*, **1**(1), 61–7.
50. FRANK, M. (1973). *J. Gen. Physiol.*, **61**, 588–618.
51. ERICKSON, R. P. (1963). In: *Olfaction and Taste*. Proceedings of the 1st International Symposium. Wenner-Gren Center, Ed. Zotterman, Y. Macmillan, New York, 205–13.
52. ERICKSON, R. P., DOETSCH, G. S. and MARSHALL, D. A. (1965). *J. Gen. Physiol.*, **49**, 247–63.
53. DOETSCH, G. S., GANCHROW, J. J., NELSON, L. M., and ERICKSON, R. P. (1969). In: *Olfaction and Taste*. Ed. Pfaffmann, C. Rockefeller University Press, New York, 492–511.
54. DOETSCH, G. S. and ERICKSON, R. P. (1970). *J. Neurophysiol.*, **33**(4), 490–507.
55. SCOTT, T. R. Jr. and ERICKSON, R. P. (1968). *J. Neurophysiol.*, **34**, 868–84.
56. ERICKSON, R. P., COVEY, E. and DOETSCH, G. S. (1980). *Brain Research*, **196**, 513–19.
57. PFAFFMANN, C., FRANK, M. and NORGREN, R. (1979). *Ann. Rev. Psychol.*, **30**, 283–325.
58. WOOLSTON, D. C. and ERICKSON, R. P. (1979). *J. Neurophysiol.*, **42**(5), 1390–1409.
59. HELLEKANT, G., GLASER, D., BROUWER, J. N. and VAN DER WEL, H. (1976). *Acta Physiol. Scand.*, **97**, 241–50.
60. OGAWA, H., SATO, M. and YAMASHITA, S. (1968). *J. Physiol.* **199**, 223–40.
61. ARISTOTLE. *De Anima*, Book II, Ed. Ross, Sir William David, Oxford Clarendon Press, (1961). Ch. 10, 422b10–15.
62. LINNAEUS (1751). As quoted by Henning.[64]
63. CHEVREUL, M. E. (1824). *Annales de Chimie et de Physique*, **26**, 386–90.
64. HENNING, H. (1916). *Zeitschrift für Psychologie mit Zeitschrift für Angewandte Psychologie und Charakterkunde*. **74**, 203–19.
65. FICK, A. (1864). Anatomie des Geschmacksorganes. In: *Lehrbuch der Anatomie und Physiologie der Sinnesorgane*, M. Schauenburg and Company, Lahr, 67–88.
66. YOSHIDA, M. and SAITO, S. (1969). *Jap. Psychol. Res.*, **11**(4), 149–66.
67. YOSHIDA, M. (1963). *Jap. J. Psychol.*, **34**(1), 25–35.
68. MCBURNEY, D. H. (1974). *Chem. Senses Flavor*, **1**, 17–28.
69. MCBURNEY, D. H. and GENT, J. F. (1979). *Psychol. Bull.*, **86**, 151–67.
70. SCHIFFMAN, S. S. and ERICKSON, R. P. (1971). *Physiol. Behav.*, **7**, 617–33.
71. SCHIFFMAN, S. S. and DACKIS, C. (1975). *Perception and Psychophysics*, **17**(2), 140–6.
72. SCHIFFMAN, S. S., MC ELROY, R. D. and ERICKSON, R. P. (1980). *Physiol. Behav.*, **7**(2), 140–6.

73. ERICKSON, R. P. (1977). In: *Olfaction and Taste VI*, Eds Le Magnen, J. and MacLeod, P., IRL London, 369–76.
74. ERICKSON, R. P. and COVEY, E. (1980). *Physiol. Behav.*, **25**, 527–33.
75. FAURION, A. SAITO, S. and MACLEOD, P. (1977). In: *Olfaction and Taste VI*, Eds Le Magnen, J. and MacLeod, P., IRL, London, 60.
76. FAURION, A., SAITO, S. and MACLEOD, P. (1980). *Chemical Senses*, **5**(2), 107–21.
77. BENZECRI, J. -P. et al. (1973). *L'analyse des Données. Tome 2, L'analyse des Correspondances*, Dunod, Paris.
78. SCHIFFMAN, S. S., REILLY, D. A. and CLARK, T. B. III (1979). *Physiology and Behavior*, **23**, 1–9.
79. FAURION, A., BONAVENTURE, L., BERTRAND, B. and MACLEOD, P. (1980). In: *Olfaction and Taste, VII*, Ed. Van der Starre, H., IRL, London, 86.
80. MILLER, I. J. JR. (1971). *J. Gen. Physiol.*, **57**, 1–25.
81. OAKLEY, B. (1975). *Chemical Senses and Flavor*, **1**(4), 431–42.
82. CLAVERIE, P. (1978). In: *Intermolecular Interactions: from Diatomics to Biopolymers*, Ed. Pullman, B. Wiley, New York, 69–305.
83. PFAFFMANN, C. (1955). *J. Neurophysiol.*, **18**, 429–40.

15

Synergism and the Sweet Response

G. G. BIRCH, G. OGUNMOYELA and S. L. MUNTON
*National College of Food Technology,
University of Reading,
Weybridge, Surrey, UK*

ABSTRACT

Synergism in sugar/sugar mixtures or sugar/artificial sweetener mixtures may be positive (enhancement) or negative (suppression) and the taste response to a defined mixture contains much information about the sugar/receptor interaction mechanism. Careful analysis of intensity/time data obtained when panellists taste these mixtures reveals the complete dominance of one molecular species at the receptor and suggests a basis for the elucidation of accession efficiencies of sapid molecules to receptors. This approach may allow a distinction to be made between the accession of a sweet molecule to a receptor and its intrinsic activity after accession. Thus synergism, taste modification and the entire process of taste chemoreception may be explained by a molecular model of the taste events.

Experiments carried out to modify the accession of sweet molecules by use of tasteless food additives show that the sweetness of certain food systems (e.g. cocoa) may be enhanced if surface tension is decreased. This type of sweetness enhancement embraces both intensity and persistence of response and seems to involve solute interactions as well as accession to receptors. It is anticipated that physicochemical investigations will eventually lead to a molecular understanding of all these phenomena.

INTRODUCTION

Synergism is a well-known phenomenon in many fields of scientific investigation and it is therefore not surprising that it has been reported in the study of taste. In mixtures of artificial sweeteners synergism is said to be responsible for the depression of 'off-flavours' and bitterness, but synergism in sweetness itself may be difficult to predict or indeed may not occur at all. Cameron[1] has reported synergism in many sugar mixtures while Hyvönen has demonstrated the phenomenon[2] in mixtures of saccharin with either fructose or xylitol.[3] On the other hand McBurney and Bartoshuk[4] have emphasised the analytic character of the primary tastes and the unlikelihood of sweet stimulus molecule interactions in mixtures. Bartoshuk[5] points out that the power function ($S = kc^n$) of taste response, S, against concentration of stimulus, c, shows how a stimulus molecule adds to itself and that synergism can be predicted from a knowledge of the power functions of the components of mixtures. Bartoshuk's conclusion thus implies that, mixing of stimuli whose exponents, n, are less than one, must result in negative synergism (i.e. depression of taste) and rules out intermolecular reactions between stimuli. All polarimetric measurements of sugar mixtures support the idea that no stimulus/stimulus interactions occur in either mixtures of different sugars or tautomeric equilibria of a single sugar.[6,7] Thus synergism of sugar mixtures (either positive or negative) must, if it exists, involve sugar/water interactions, sugar/receptor interactions, neural mechanisms or combinations of all three possibilities. Unless these mechanisms can be distinguished and one or more of them eliminated by experimental evidence, the phenomenon of synergism will not be explained in the context of a suitable model of taste chemoreception.

BITTER/SWEET INTERACTIONS

McBurney and Bartoshuk have reported[4] the effectiveness of adaptation to one basic taste in modifying the other basic tastes; adaptation to sucrose, for example, causes quinine to taste more bitter. On the other hand, tasting quinine and sucrose in a mixture[8] lowers the overall taste intensity because the primary tastes mutually depress one another. Thus adaptation and stimulus/stimulus interaction seem to cause opposite effects and clearly, temporal factors (e.g. administration time of stimulus and persistence of taste experience) will govern the overall nature of the taste response.

Lawless[9] reported evidence that taste suppression in bitter/sweet mixtures was due to neural inhibition rather than solute interactions or competition for common receptors. This was based on pretreatment with gymnemic acid or pre-adaption with sucrose which lowered the perception of sweetness in a bitter/sweet mixture and consequently nullified the suppression of bitterness. His results certainly do seem best explained by neural inhibition but the lack of competition of sweet and bitter molecules for a common receptor reported by him had already been established in previous studies of sucrose and quinine mixtures.[10] Furthermore, Lawless did not consider the likelihood of sugar/water competition or sugar/gymnemic acid competition for a common receptor, and the proximity of discrete sweet and bitter receptor sites therefore still remains as a distinct possibility[10] to account for the taste of bitter/sweet molecules.

INTENSITY/TIME ANALYSIS AS A CLUE TO SYNERGISM

It has recently been pointed out[11-15] that intensity/time profiles of a sweet response can give valuable information about sweet receptor function and that therefore the kinetics of sweet taste response demand a kinetic explanation of stimulus/receptor interaction. In other words the rate theory of stimulus/receptor interaction, which implies a flux of stimulus molecules interacting with receptors, has to be adopted to explain the intensity/time behaviour. A device for improved recording of these data (Sensory Measuring Unit for Recording Flux, 'SMURF') has recently been described[16] after studies at Weybridge. The SMURF has allowed us to demonstrate simple power functions for both intensity and persistence of response in the sweetness of several sugars and sugar alcohols. Thus when simple binary mixtures of any *two* of these sweet stimuli (in various proportions) are tasted, the intensity/time response of the mixture can be interpreted by the combined power functions of the components to elucidate the dominant substances in the taste response.[17] Simple calculations along these lines are not always arithmetically possible but those that are show that one substance in any binary mixture is always absolutely dominant. In other words each binary mixture behaves as if it consists of only one of its components and the 'effective concentration' of the absolutely dominant component can be calculated (Table 1).

TABLE 1
CONCENTRATIONS AND EFFECTIVE CONCENTRATIONS OF BINARY MIXTURES BY INTENSITY/TIME SWEETNESS ANALYSIS

Mixture	Concentrations (% w/v)	'Effective concentrations' (% w/v)
Sucrose: sorbitol	6·9:3·7	<0·01:13·9
Sucrose: xylose	6·9:3·2	<0·01:11·4
Sucrose: maltose	6·9:5·4	<0·01:20·9
Sucrose: galactose	6·9:4·3	<0·01:14·6
Glucose: galactose	11·3:4·3	<0·11:13·9

This observation means that synergism in mixtures is certainly not predictable from conventional power functions of individual components. Table 1 in fact shows that the dominant component may have an 'effective concentration' greater than the concentration which is presented. This is not anomalous, but simply an illustration of the concept of accessibility of receptors. Thus the presence of one substance may influence the fraction of presented molecules of a second substance that successfully recruits receptors. The 'effective concentrations' presented in Table 1 may therefore represent real concentrations of stimuli in the localised concentration of the receptor site.

Previously Yamaguchi et al.[18] had also demonstrated a degree of dominance in mixtures. However, work in the field of synergism has traditionally ignored the important time factor in the sweet response and the accessibility characteristics of receptors to the different sweet stimuli have hitherto remained obscure. Table 1 therefore presents the first set of data related to the accession efficiencies[13] of sapid molecules made available by psychophysical studies.

It is worth noting in Table 1 that two of the dominant molecular species (maltose and sorbitol) are substances already identified by Schiffman et al.[19] as being similar in taste profile and different from other simple sweeteners; this raises several more questions about the sweetening power of individual molecules. What are the structural reasons why one molecule should gain preferential accession to the receptor? The question might be more easily answered when comparing a sugar with, say, an artificial sweetener such as a saccharin, when the lipophilic characteristics of the latter offer an explanation of its favourable partition between the oral fluid and the hydrophobic surface of the taste cell membrane. However, in the sugars and their simple derivatives, there are no lipophilic characteristics whatsoever and differences in their accession efficiencies to receptors must be due to their intrinsic stereochemistry and hydrogen-bonding capacities.

SUGAR/WATER INTERACTION AND ITS SIGNIFICANCE IN SYNERGISM

If one substance has the property of modifying the taste of another substance it seems logical that the intermediation of water molecules should be involved. Indeed the accession of a stimulus molecule to a receptor probably involves competition with water and herein lies a possible explanation of the receptor accession efficiencies. The ability of, say, a sugar molecule to displace water molecules from a receptor, probably depends on its degree of hydration and the strength of its hydrated complex. These are of course in turn governed by the structure of the molecule and its hydrogen-bonding capacity. It is noticeable from Table 1 that several of the dominant molecular species (e.g. sorbitol, maltose, galactose) are those of lower intrinsic sweetness. Such molecules are capable of extensive intramolecular hydrogen bonding which may depress their capacity for hydration. This distinguishes them from sucrose which is fully hydrated at concentrations (e.g. 5%) commonly employed in taste-experiments. All studies of synergism in mixtures of sweeteners should therefore be regarded as investigations of competition between sweet molecules and water molecules and this of course re-introduces McBurney and Bartoshuk's concept[4] of water taste as an explanation of adaptation phenomena. No study of synergism can be properly understood unless the dynamic sequence of events underlying the taste experience is interpretable at the molecular level.

ACCESSION EFFICIENCY, LOCALISED CONCENTRATION AND SYNERGISM

Although accession efficiency must be a primary determinant of sweetness intensity and persistence, the function of a molecule after accession to the receptor must also contribute to the response. Recently[11,12] a model embodying localised concentration and ionophore trigger efficiency has been advanced, the former providing an organised 'store' of stimuli giving rise to persistence of taste. The idea of localised concentration is not new and has in fact been established in isolated insect chemoreceptors.[20] Such a model therefore offers a convenient explanation of the intense sweetness and persistence of thaumatin and related protein molecules.[15] On the basis of the localised concentration concept it is easy to envisage that the concentration of stimulus molecules at a receptor could exceed that which is presented and this of course infers a variety of approaches to the creation of synergism by chemical manipulation. Any substance which might affect the accession efficiency, localised concentration or ionophore trigger efficiency of a sapid molecule should affect the nature of its taste response, determinable by its intensity/time profile. There are of course multitudes of possible candidates for these diverse effects and a good starting point in any search for sweetness enhancers would be lists of permitted food additives currently in use throughout the world.

SURFACTANTS AND SYNERGISM

A logical starting point in the search for sweetness enhancers is the class of additives known as surfactants. This type of molecule can modify the environment of the receptor, facilitating the partition of sapid molecule from the oral fluid to the taste cell membrane. It might also complex with a sweet molecule such as sucrose and facilitate its diffusion, as a complex, to the receptor. Moreover, such a complex might alter the conformation of the sugar and thus modify the effectiveness of its primary AH,B system. Complex formation of this type has been demonstrated[14] between sucrose and either of the surfactants glycerol monostearate or lecithin (Fig. 1). If either of these surfactants is tested for its gustatory effect in a typical food system (cocoa) it is seen to elevate both the intensity and persistence of sweetness response[14] and indeed a weak sucrose solution may be made to taste more sweet than a strong sucrose solution by interaction with a tasteless surfactant in this way (Table 2).

It is well known that the specific rotation of sugars can be altered by concentration and the presence of additives and this allows a molecular

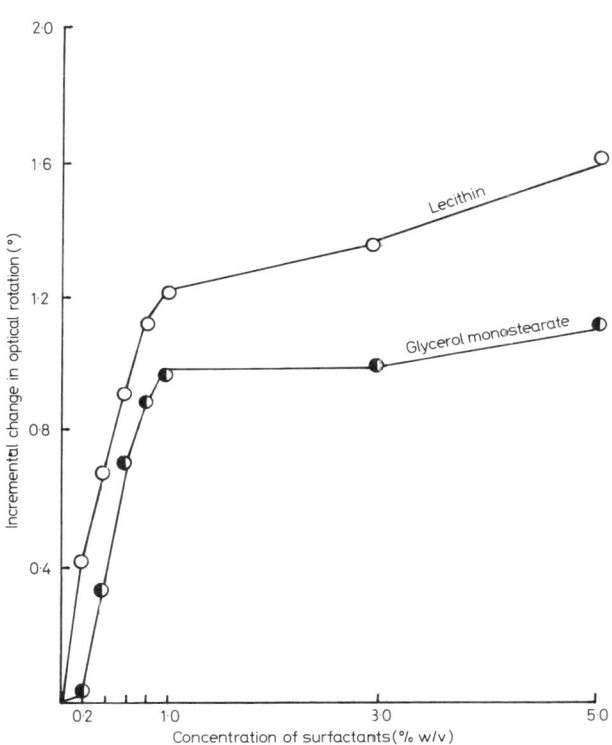

FIG. 1. Interaction of surfactants with 1% w/v sucrose solution.

interpretation of the resulting taste changes as well as any ensuing nutritional implications. Previous attempts to use surfactants to modify gustatory effects have resulted in conflicting literature reports.[21,22] This is presumably because the class of surfactants includes hydrocolloids which modify viscosity and other rheological properties of sugar solutions in addition to modifying surface tension. Different combinations of physicochemical properties in the different surfactants therefore give rise to different (sometimes opposite) gustatory effects which may further vary with concentration.[23] Although both glycerol monostearate and lecithin elevate sweetness and both depress surface tension, only glycerol monostearate elevates viscosity.[24] Thus the surface tension depression seems to be of significance in the enhancement of the taste response (Table 3). Further examples of synergism in complex food systems have recently been reported.[25]

TABLE 2
ENHANCEMENT OF SWEETNESS INTENSITY AND SWEETNESS PERSISTENCE BY SURFACTANTS IN COCOA

Concentration sucrose added (% w/v)	Sweetness intensity (SMURF units)							Sweetness persistence (s)						
		Glycerol monostearate added (% w/v)			Lecithin added (% w/v)				Glycerol monostearate added (% w/v)			Lecithin added (% w/v)		
	0	0.5	1.0	0	0.5	1.0		0	0.5	1.0	0	0.5	1.0	
10	6.5	7.3	7.9	6.5	7.2	7.8		24.3	26.8	28.5	24.4	25.2	27.8	
20	7.5	8.3	8.5	7.5	7.7	8.2		24.6	36.0	36.5	24.6	31.6	32.0	
30	8.0	8.4	8.9	8.0	8.1	8.6		26.6	36.7	39.8	26.6	34.8	39.5	

TABLE 3
APPARENT VISCOSITIES AND STATIC SURFACE TENSIONS OF SWEETENED COCOA DRINKS WITH SURFACTANTS

Concentration sucrose added (% w/v)	Apparent viscosity (Brookfield cP)							Static surface tension (mN/m)						
		Glycerol monostearate added (% w/v)			Lecithin added (% w/v)				Glycerol monostearate added (% w/v)			Lecithin added (% w/v)		
	0	0.5	1.0	0	0.5	1.0		0	0.5	1.0	0	0.5	1.0
10	15.0	35.0	77.5	15.0	15.0	17.5		59.5	51.2	50.7	59.5	45.4	44.3
20	20.0	85.0	97.5	20.0	22.5	25.0		59.6	51.7	50.9	59.6	45.4	44.1
30	22.5	90.0	113	22.5	25.0	27.5		59.8	52.3	51.1	59.8	45.1	43.8

These examples of sweetness enhancement are included under synergism as they are considered to be the same in principle as those involving synergism of sugar mixtures. In other words solute/solute interactions, solute/water interactions and modification of receptor environment underlie all the observations of synergism and enhancement and the psychophysical analysis of the effects by intensity/time profiling allows a deeper understanding of the ways these effects are manifested. Enhancement of sweetness intensity signifies the operation of greater numbers of receptors whereas enhancement of persistence signifies greater numbers of stimulus molecules at localised concentration sites. The shapes of the intensity/time curves themselves offer information about the rate of accession to receptors and hence about the flux of stimuli giving rise to gustatory effects.

CONCLUSION

Although synergism is of practical significance in food science and technology, its real importance lies in its theoretical contribution to an understanding of taste chemoreception. Research on the nature of the sweet glycophore has revealed structural differences between sugars and other sweet stimuli which might determine their interaction with the sweet receptor. These differences can be elucidated by a study of synergism in mixtures in order to distinguish between the affinity (accession efficiency) and intrinsic activity (ionophore trigger efficiency) of the different sweet stimuli. Experimental evidence suggests that solute/solute interactions, solute/water interactions and solute/receptor interactions may all contribute to observable synergistic effects and the data support the concept of a two-stage chemoreception process involved in the taste response. Synergism is therefore viewed as an aid to the molecular interpretation of the sweet taste phenomenon.

REFERENCES

1. CAMERON, A. (1947). *The Taste Sense and the Relative Sweetness of Sugars and other Sweet Substances*, Sugar Research Foundation (NY) Scientific Report Series. 9. 72 pp.
2. HYVÖNEN, L., KURKELA, R., KOIVISTOINEN, P. and RATILAINEN, A. (1978). *J. Fd Sci.*, **43**, 251–4.

3. HYVÖNEN, L. (1980). *Varying Relative Sweetness*, EKT Sarja Series 546 University of Helsinki.
4. MCBURNEY, D. and BARTOSHUK, L. M. (1973). *Physiol. Behaviour*, **10**, 1101–6.
5. BARTOSHUK, L. M. (1977). In: *Sensory Properties of Foods*, Eds Birch, G. G. and Parker, K. J., Applied Science Publishers, London, 1–26.
6. SHALLENBERGER, R. S., BRAVERMAN, S. E., and GUILD, W. E. Jr., (1980). *Fd Chem.* **5**, 207–16.
7. HYVÖNEN, L., VARO, P. and KOIVISTOINEN, P. (1977). *J. Fd Sci.*, **42**, 657–9.
8. BIRCH, G. G., COWELL, N. D. and YOUNG, R. (1972). *J. Sci. Fd Agric.*, **23**, 1207–12.
9. LAWLESS, H. (1979). *J. Comp. Physiol. Psychol.*, **93**, 538–47.
10. BIRCH, G. G. and MYLVAGANAM, A. R. (1976). *Nature*, **260**, 632–4.
11. BIRCH, G. G. (1979). In: *Developments in Food Science*, 2. Eds Chiba, H., Fujimaki, M., Iwai, K., Mitsuda, H., and Morita Y. Elsevier, Amsterdam, 367–72.
12. BIRCH, G. G. (1980). In: *Carbohydrate Sweeteners in Foods in Nutrition*, Eds Koivistoinen, P. and Hyvönen, L. Academic Press, New York, 61–75.
13. BIRCH, G. G. (1981). In: *Biochemistry of Taste and Olfaction*, Eds Cagan, R. and Kare, M. Academic Press, New York, 163–73.
14. BIRCH, G. G. and OGUNMOYELA, G. (1980). *J. Fd Sci.*, **45**, 981–4.
15. BIRCH, G. G., LATYMER, Z. and HOLLAWAY, M. (1980). *Chem. Senses*, **5**, 63–78.
16. BIRCH, G. G. and MUNTON, S. L. (1981). *Chem. Senses*, **6**, 45.
17. MUNTON, S. L. and BIRCH, G. G. (1981). Unpublished results.
18. YAMAGUCHI, S., YOSHIKAWA, T., IKEDA, S. and NINOMIYA, T. (1970). *Agr. Biol. Chem.*, **34**, 181–6.
19. SCHIFFMAN, S. S., REILLY, D. A. and CLARK, T. E. (1979). *Physiol. Behav.*, **23**, 1–9.
20. HANNSEN, K. (1978). In: *Taxis and Behaviour. Receptors and Recognition*, Series B. Vol. 5. Ed. Hazelbauer, G. L. Chapman and Hall. London, 233–92.
21. PANGBORN, R. M., GIBBS, Z. M. and TASSAN, C. (1978). *J. Text., Studies*, **9**, 415–20.
22. STONE, H. and OLIVER, S. M. (1966). *J. Fd Sci.*, **31**, 129–33.
23. PAULUS, K. and HAAS, E. M. (1980). *Chem. Senses*, **5**, 23–32.
24. OGUNMOYELA, G. and BIRCH, G. G (1982). *J. Agr. Fd Chem.*, (in press).
25. DIEDRICHS, F., BIRCH, G. G. and MUNTON, S. L. (1980). In: *Inter-Eis 80. ZDS*, International Confectionery Institute. Solingen. W. Germany. 1–24.

16

Multidimensional Concepts in Sweetness Evaluation

SUSAN S. SCHIFFMAN
*Department of Psychiatry,
Duke Medical Center,
Durham, North Carolina, USA*

ABSTRACT

This paper describes two studies that demonstrate how the mathematical technique of multidimensional scaling can be used to delineate the psychophysical properties of sweeteners and to explore the possible number of receptor sites mediating sweetness. In the first study, 17 sweeteners varying widely in chemical structure were arranged in a three-dimensional space on the basis of experimental measures of similarity. The sugars (fructose, glucose, sorbose, xylitol and xylose) were placed near one another, and those with a syrupy component (sorbitol and maltose) fell close to this group. Aspartame and calcium cyclamate were the two artificial sweeteners that were found to be closest in quality to the sugars. Stimuli with long aftertastes (neohesperidin dihydrochalcone, monellin, and thaumatin) were positioned away from these. Stimuli with the highest bitter and metallic ratings (acetosulfam, sodium saccharin, rebaudioside, stevioside along with D-tryptophan) also tended to group together. Individual spaces revealed that sodium saccharin and acetosulfam were arranged slightly closer to the sugars in spaces for tasters of PTC than for non-tasters. In the second study, the method of cross adaptation was used to investigate the receptor properties of sweetness. It was assumed that if adaptation to one stimulus reduced the sensory response to another, then this may indicate that the two stimuli share a common receptor site. Both adaptation and enhancement were found in this study. Implications for the possible number of receptor sites are discussed.

INTRODUCTION

The purpose of this paper is to demonstrate how the mathematical technique called multidimensional scaling can be used to gain a greater understanding of the nature of the sweet taste. Multidimensional scaling (MDS) is a computer-based procedure that can be used to systematise data when the organising concepts are not well understood. This is particularly advantageous in examining sweetness because at the present time it is not possible to predict, from the chemical structure, whether a molecule will taste sweet. Multidimensional scaling represents the similarities among objects in geometrical form as in a map. A set of numbers that expresses all or most combinations of similarities is utilised by MDS to represent objects judged to be experimentally similar to one another, close to one another in a resultant spatial map. Objects judged to be dissimilar are arranged distant from one another.

Two examples will be given here that illustrate how multidimensional scaling can be used to gain a greater understanding of (1) the perceptual range of sweetness, and (2) the underlying receptor mechanisms involved in mediating the sweet taste.

EXPERIMENT 1

The purpose of this experiment was to delineate the perceptual differences among sweeteners utilising multidimensional scaling techniques.[1] Twelve subjects rated the similarities among 17 sweeteners shown in Table 1 at the concentrations given. The chemical structures are illustrated in Fig. 1. When pretested the concentrations used were found to be approximately equal in overall psychophysical intensity. Each subject made 136 judgments (C_2^{17}) during 20 1-h sessions. Interpair intervals were at least 5 min in duration (and sometimes as long as 10 min or more when thaumatin and monellin were used as stimuli). The ratings of similarity were made by marking an X along a 5 in line connecting the words 'same' and 'different'.

When all the judgements of similarity were completed, ratings were made on a series of adjective scales which had been used previously.[2-6] Four additional adjectives were included: syrupy,

TABLE 1
STIMULI: CLASSIFICATION, SOURCE, CONCENTRATIONS, AND USE IN EXPERIMENTS

Compound	Classification	Source	Concentration used	Expt 1	Expt 2 Pt 1	Expt 2 Pts 2, 3
Acetosulfam	Oxathiazinone dioxide (methyl derivative); 3, 4 dihydro-6-methyl-1, 2, 3 oxathiazin-4-one-2, 2-dioxide potassium salt	Hoechst (German)	0·35%	X	X	X
Aspartame	Dipeptide: L-aspartyl-L-phenylalanine methyl ester	Searle	0·25%	X		X
Ca cyclamate	Calcium cyclohexyl sulfamate	Monsanto	0·6%	X	X	X
Fructose	Monosaccharide ketohexose	Sigma	0·6M	X	X	
Galactose	Monosaccharide aldohexose	Sigma	1·0M		X	
Glucose	Monosaccharide aldohexose	Sigma	1·1M	X	X	X
Maltose	Disaccharide	Sigma	1·2M	X	X	
Monellin	Protein (MW 10 700)	Worthington and R. Cagan	0·025%	X		
Neohesperidin dihydrochalcone	Dihydrochalcone glycoside	California Aromatics	0·016%	X		X
Rebaudioside A	Diterpene glycoside	Ajinomoto	0·07%	X	X	
Saccharin (sodium salt)	o-sulfobenzimide: 1, 2-benzo-thiazol-3(2H)-one-1, 1-dioxide, Na⁺ salt	Logica International	0·045%	X	X	X
Sorbose	Monosaccharide ketohexose	Sigma	1·2M	X		
Sorbitol	Polyhydric alcohol	Sigma	1·2M	X	X	
Stevioside	Diterpene glycoside	Ajinomoto	0·09%	X		
Thaumatin	Several distinct proteins (MW 18 000–21 000)	Tate and Lyle Ltd	0·01%	X		
D-tryptophan	D-amino acid	Sigma	0·3%	X		X
Xylitol	Polyhydric alcohol	Sigma	1·0M	X	X	X
Xylose	Monosaccharide aldopentose	Sigma	1·3M	X	X	

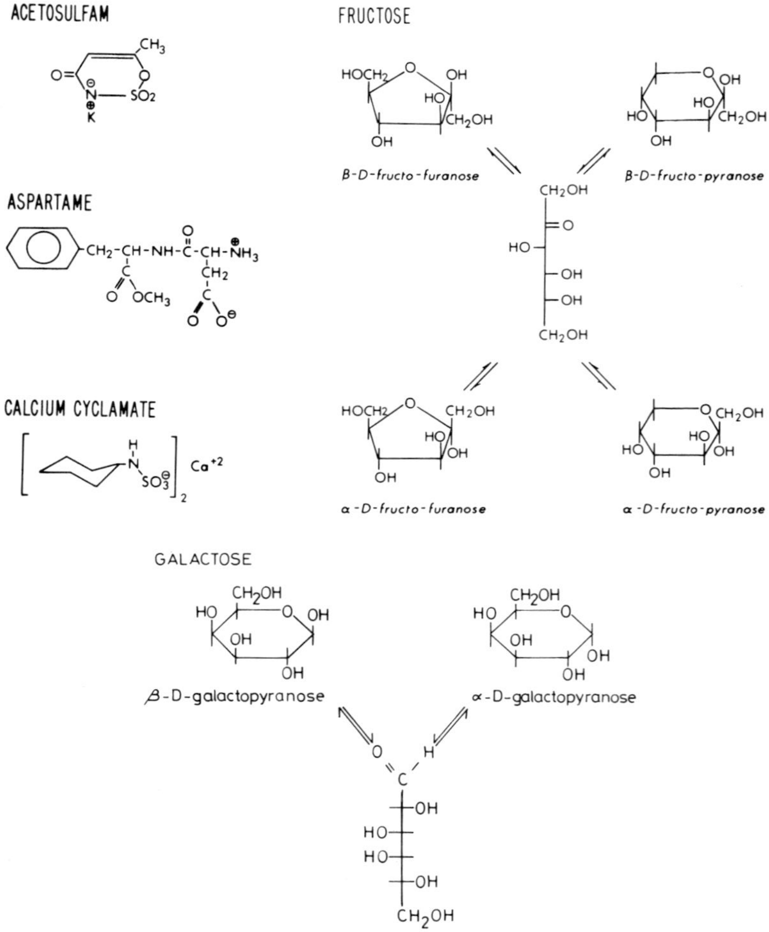

FIG. 1.

taste fades fast, delayed sweetness, and aftertaste. Subjects were also encouraged to make additional comments.

The ratings along the 'same–different' line were transcribed from 0 to 100 and analysed by two different multidimensional scaling procedures: INDSCAL[7] and ALSCAL.[8] INDSCAL is a metric multidimensional scaling procedure that assumes experimental measures of similarity obey restrictions of either the interval or

GLUCOSE

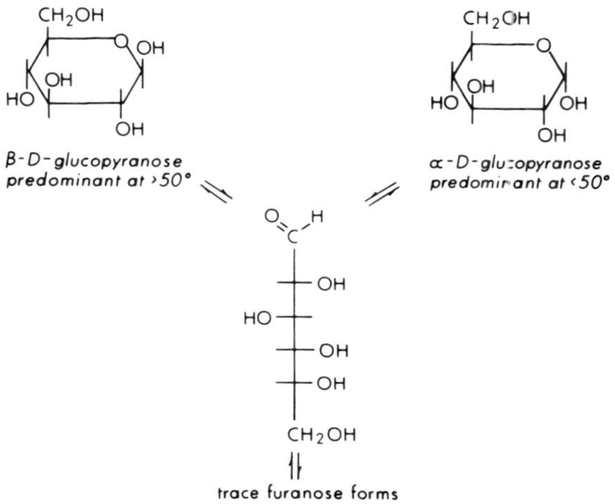

β-D-glucopyranose
predominant at >50°

α-D-glucopyranose
predominant at <50°

trace furanose forms

MALTOSE

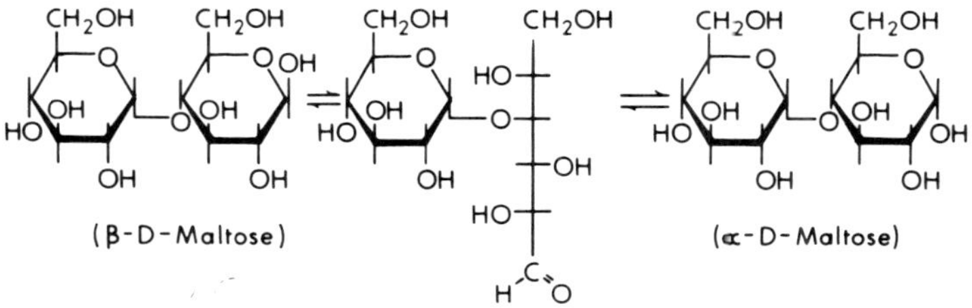

(β-D-Maltose)

(α-D-Maltose)

MONELLIN-PROTEIN

β-NEOHESPERIDIN DIHYDROCHALCONE

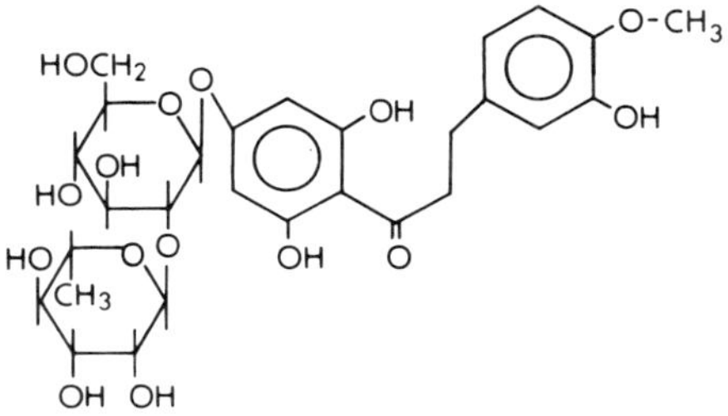

FIG. 1.—contd.

REBAUDIOSIDE

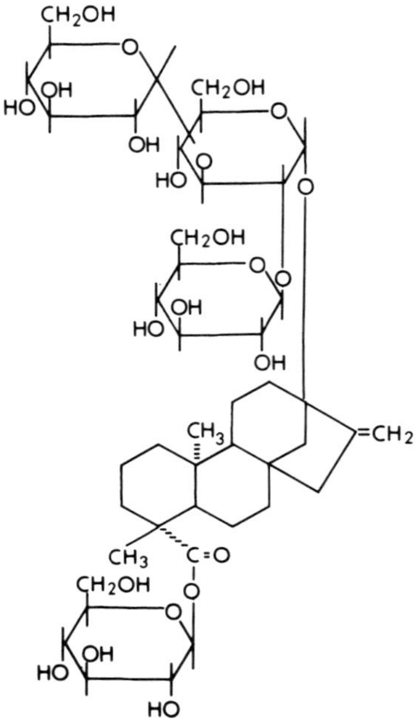

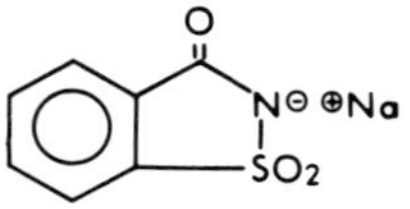

STEVIOSIDE

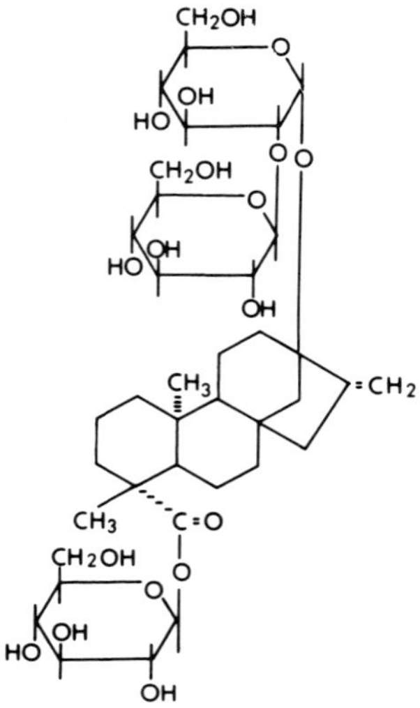

THAUMATIN-PROTEIN

D-TRYPTOPHAN

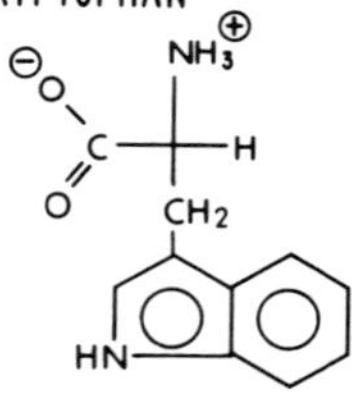

SACCHARIN (Na SALT)

FIG. 1.—contd.

ratio level of measurement. ALSCAL possesses a nonmetric option that allows treatment of data at the ordinal level of measurement such that the distances in the derived geometric representation correspond to the rank order of the similarity judgements. Both of these methodologies were employed to avoid making *a priori* assumptions about the level of measurement appropriate to the data. In addition to similarity spaces, both INDSCAL and ALSCAL provide weights for individual subjects on each of the dimensions of a similarity space common to all subjects. These weights reveal any idiosyncratic ratings by the subjects and thus

SORBITOL

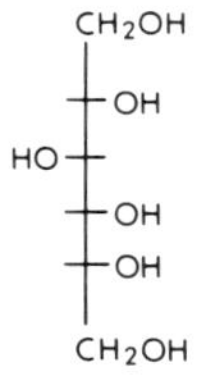

```
     CH₂OH
      |
      +—OH
      |
  HO—+
      |
      +—OH
      |
      +—OH
      |
     CH₂OH
```

XYLITOL

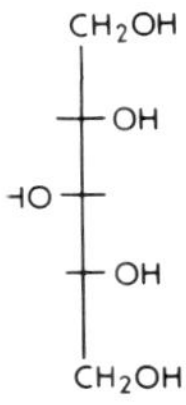

```
     CH₂OH
      |
      +—OH
      |
  HO—+
      |
      +—OH
      |
     CH₂OH
```

SORBOSE

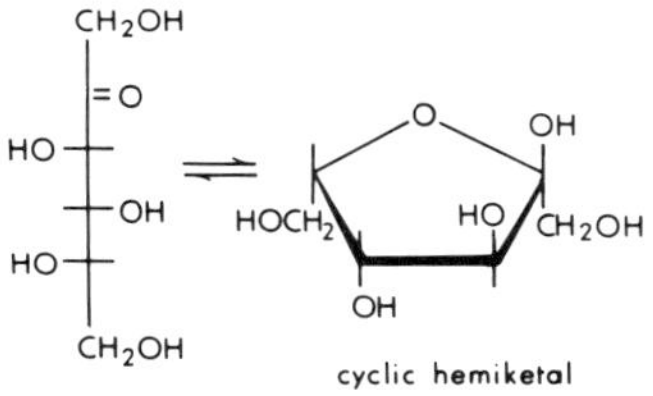

```
     CH₂OH
      |
      +=O
      |
  HO—+
      |                       O        OH
      +—OH       ⇌     HOCH₂        
  HO—+                     HO   CH₂OH
      |                        OH
     CH₂OH
                    cyclic hemiketal
```

XYLOSE

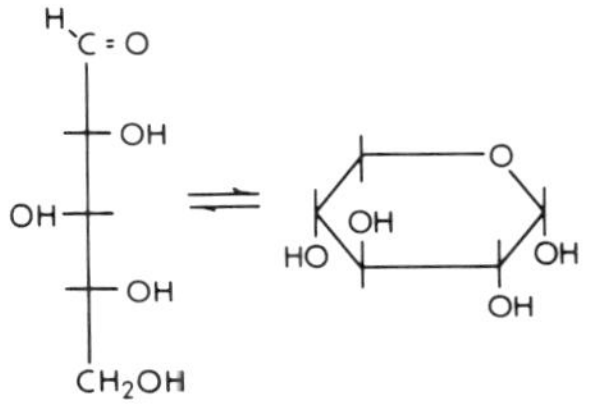

```
       H
        \
         C=O
         |
         +—OH
         |
     OH—+           ⇌              O
         |                   HO      OH
         +—OH                    OH
         |
        CH₂OH
```

FIG. 1.—contd.

expose individual differences. A procedure called PREFMAP[9] was used to project adjective ratings through the multidimensional sweetener space such that the mean adjective ratings over all subjects were represented as scale values along unit vectors.

The results of the applications of INDSCAL and ALSCAL were virtually identical. The INDSCAL space is given in Fig. 2 and two-dimensional cross-sections through this space in Figs 3(a) and 3(b). It can be seen in the upper left-hand quadrant of Fig. 3(a) that fructose, xylose, sorbose, xylitol and glucose tend to group together, with aspartame and sorbitol located nearby. The triad monellin, thaumatin, and neohesperidin dihydrochalcone are in the lower left-hand quadrant separate from the sugar area and in the lower right-hand quadrant acetosulfam, rebaudioside, stevioside, sodium saccharin and D-tryptophan are found. Calcium cyclamate is located between the acetosulfam–rebaudioside–sodium saccharin group and the sugars.

The weight spaces that reflect individual subject differences in the INDSCAL solution are shown in Figs 4(a) and 4(b). The meaning of these spaces is given as follows: Individual subjects weight the

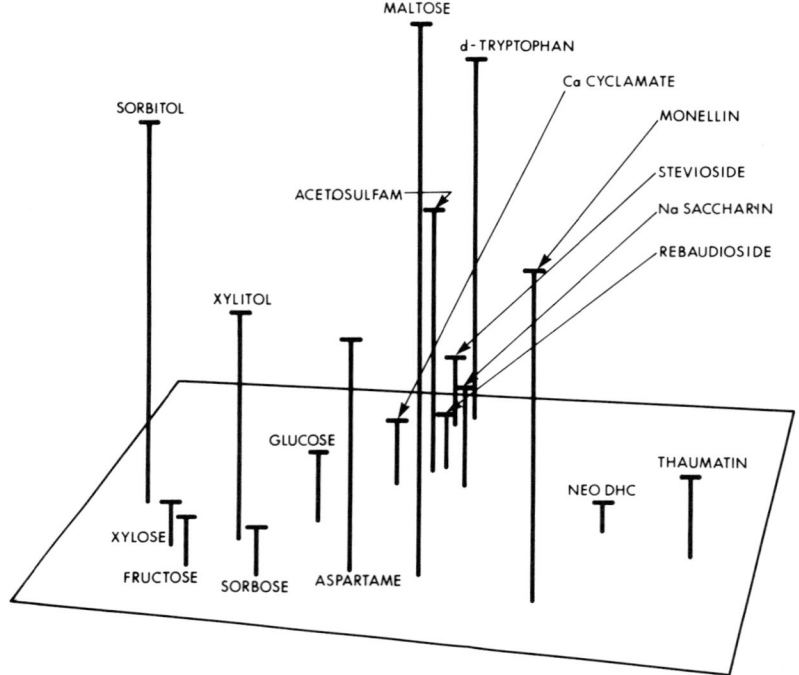

FIG. 2.

dimensions in idiosyncratic ways. For example, dimension I which is highly correlated with sweet and bitter adjective ratings was weighted more by subject 3 than dimension II. Thus, the appropriate multidimensional arrangement for this subject would be similar to that in Fig. 2 (and Fig. 3) but stretched out in an elliptical manner along the first dimension. Alternatively, for subject 8, the appropriate arrangement would be stretched out in an elliptical fashion along dimension II which is related to aftertaste. Although the weight spaces did not indicate a difference between the tasters of the bitter compound phenylthiourea (PTC) and non-tasters,[10,11] individual arrangements found using the MDS technique called SSAI[12,13] did reveal minor differences between them. There was a trend for sodium saccharin and acetosulfam to be found closer to the sugar group in individual spaces for non-tasters when compared to tasters.

Adjective ratings were also explored to determine their re-

Multidimensional Concepts in Sweetness Evaluation 295

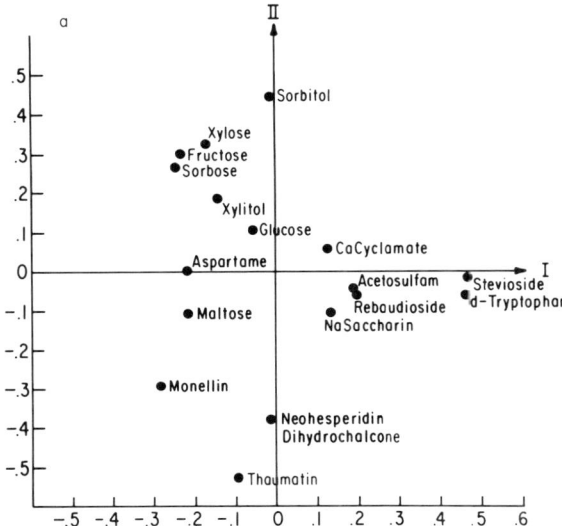

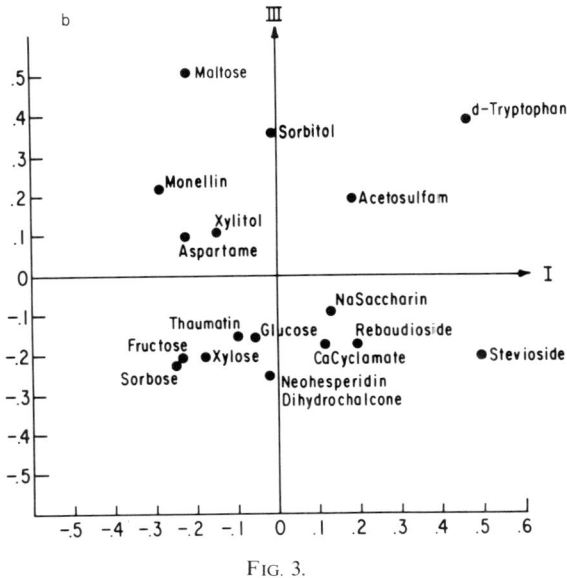

FIG. 3.

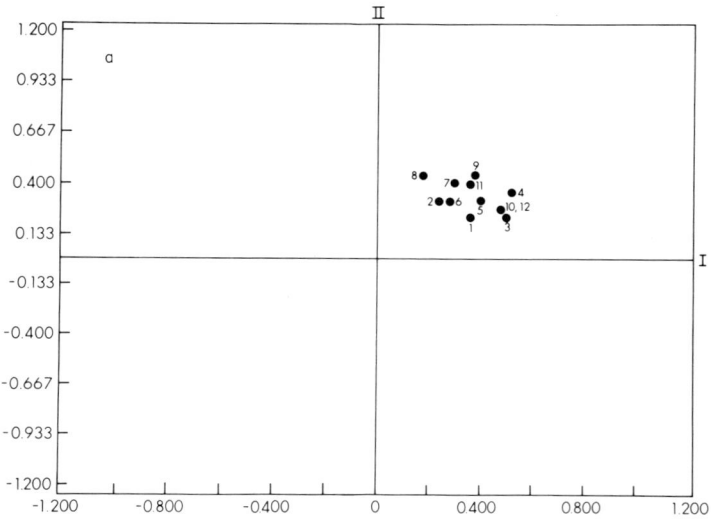

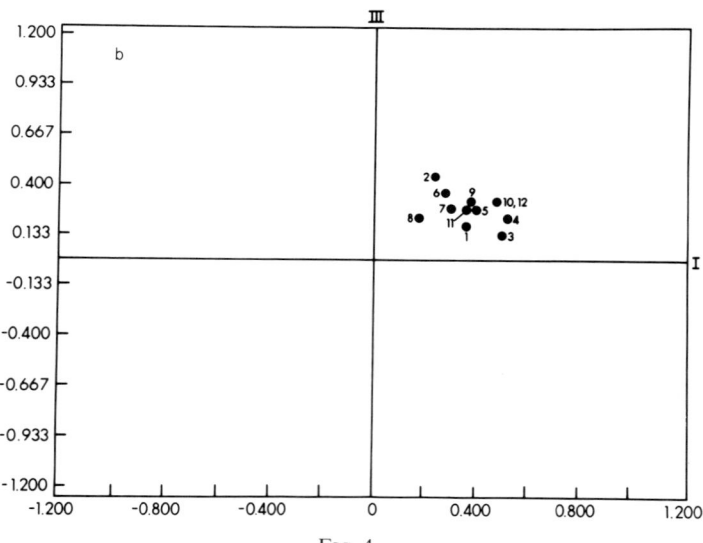

Fig. 4.

lationship to the spatial arrangement in Fig. 2 (and Fig. 3). Most of the adjective scales were not found to be relevant to the spatial arrangement. The remainder tended to reveal considerable individual variation between subjects. The variation on 10 adjective scales is shown in the histograms in Fig. 5. Circles correspond to ratings by individual subjects in given areas along the scale. Thus as an example, acetosulfam had a rating transcribed between 11–20 for one subject on the 'good' scale. Another subject's rating was transcribed between 21–30. Two subjects' ratings were transcribed from 81–90. The left-hand side of the scale corresponds to the adjective given above the histogram (e.g. 'good'; with the right-hand side corresponding to the opposite of that rating (e.g. 'bad'). It is clear that language-based adjective ratings reveal much greater variability than similarity judgements.

Although adjectives are not particularly helpful in accurately quantifying relationships among sweeteners, they can be helpful in describing the arrangements found by similarity judgements. The saccharides and polyhydric alcohols tended to be arranged close to one another in the multidimensional space, and they are rated similarly on adjective scales as relatively fast developing with less aftertaste than the other stimuli. Xylose and xylitol had slight unpleasant components which were difficult to characterise verbally. Stimuli with long aftertastes—monellin, thaumatin, and neohesperidin dihydrochalcone—tended to group together. The two artificial sweeteners, aspartame and calcium cyclamate, that were rated relatively high on the 'good' scale, tended to fall closer to the sugars than the other artificial sweeteners.

PREFMAP was applied to the geometric means of the adjective ratings in Fig. 5 to determine whether adjective scales could be projected into the similarity space. The direction cosines for adjectives that had correlations of 0·80 or above are given in Table 2.

Thus, characterisation by MDS of the perceptual differences between sweeteners reveals that of the stimuli tested here aspartame and calcium cyclamate are the best possible substitutes for the sugars because they fall closest to the sugars in the multidimensional sweetener space. The MDS approach can usefully be expanded to characterise the differences among members of a series of analogues of aspartame[14] or chlorosucroses[15] for example, to shed more light on the relationship of sweetness to chemical structure.

298 Susan S. Schiffman

FIG. 5.

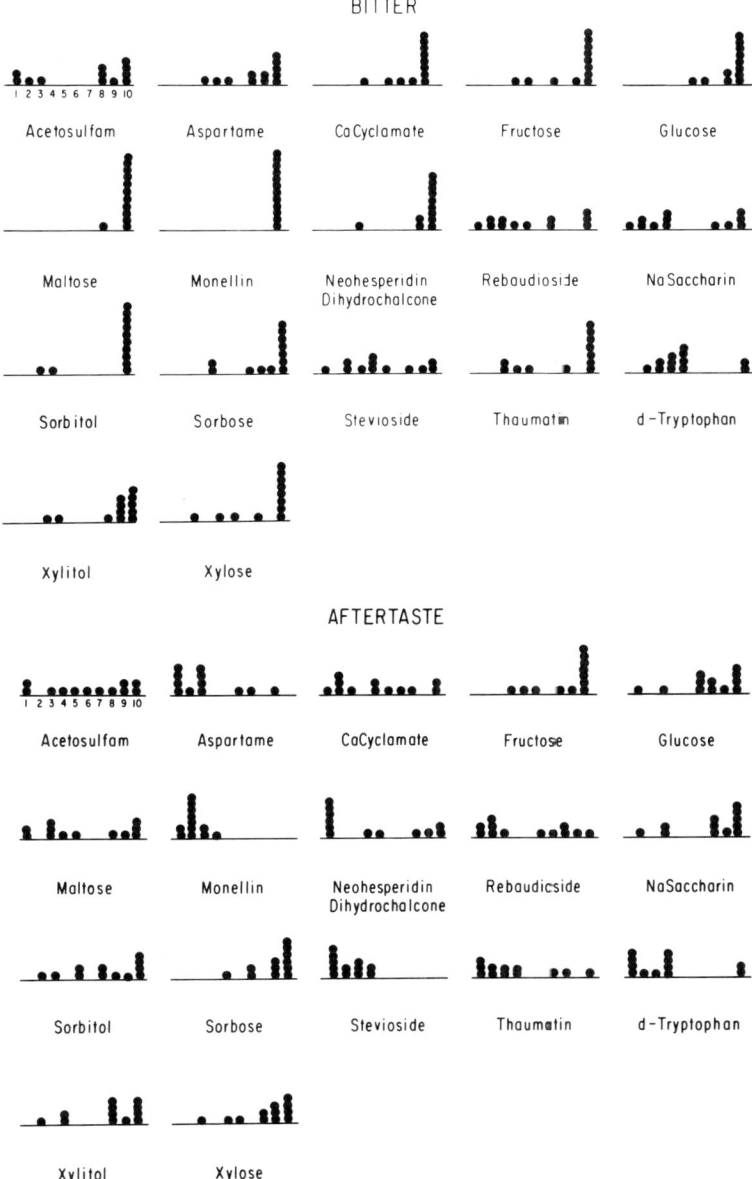

FIG. 5.—contd.

FIG. 5.—contd.

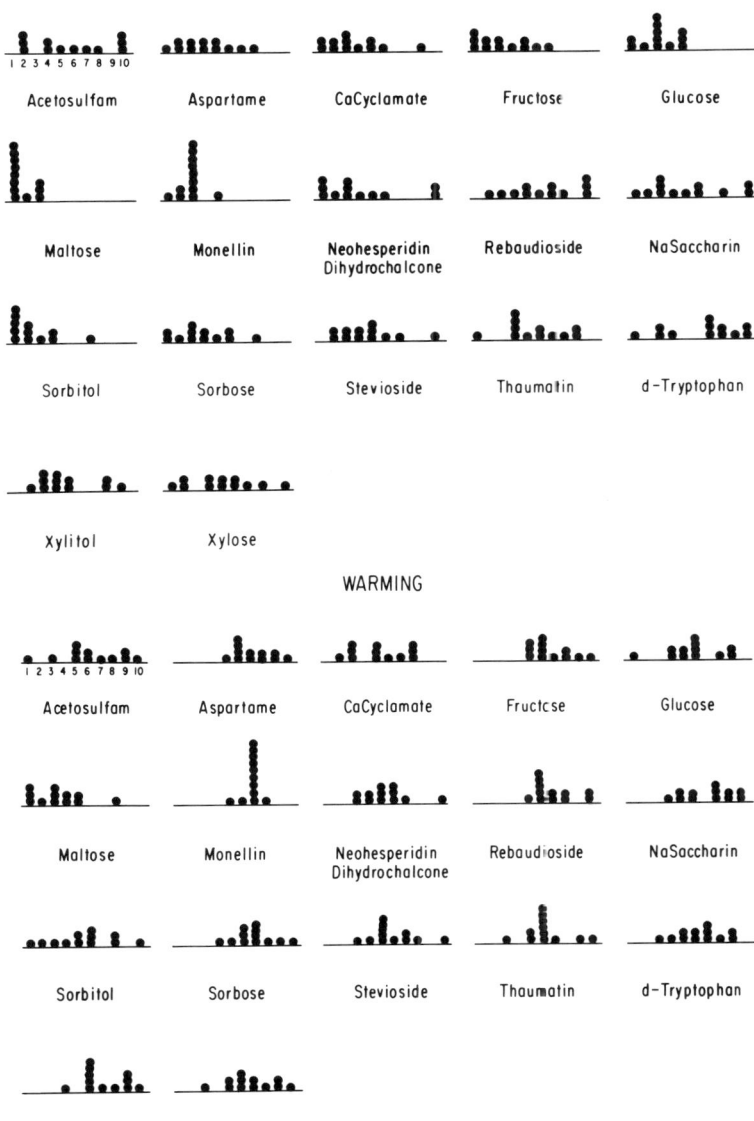

FIG. 5.—contd.

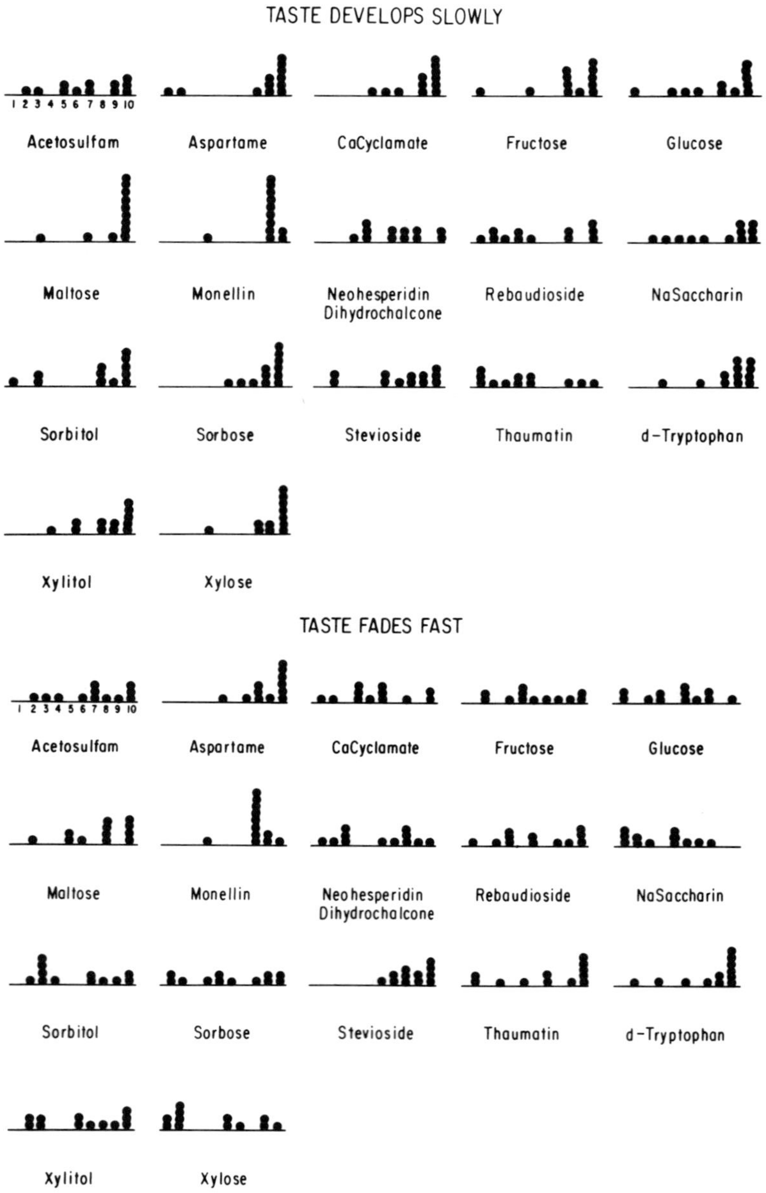
TASTE DEVELOPS SLOWLY
1 2 3 4 5 6 7 8 9 10
Acetosulfam
Aspartame
CaCyclamate
Fructose
Glucose
Maltose
Monellin
Neohesperidin Dihydrochalcone
Rebaudioside
NaSaccharin
Sorbitol
Sorbose
Stevioside
Thaumatin
d-Tryptophan
Xylitol
Xylose
TASTE FADES FAST
1 2 3 4 5 6 7 8 9 10
Acetosulfam
Aspartame
CaCyclamate
Fructose
Glucose
Maltose
Monellin
Neohesperidin Dihydrochalcone
Rebaudioside
NaSaccharin
Sorbitol
Sorbose
Stevioside
Thaumatin
d-Tryptophan
Xylitol
Xylose

FIG. 5.—*contd.*

TABLE 2

	I	II	III
Good	−0·639	0·768	0·051
Taste changes	0·424	−0·898	−0·115
Sweet	−0·832	0·409	0·375
Aftertaste	0·227	−0·933	0·281

EXPERIMENT 2

In this experiment,[16] the psychophysical method of cross adaptation was used to investigate the possible number of gustatory receptor sites for sweetness. Two stimuli may share common receptor sites if adaptation to one stimulus results in a decreased sensory response to another. Alternatively, if a decreased sensation to another stimulus does not occur after adaptation to another stimulus, a possible implication is that different receptor sites are involved.

This study consisted of three parts. In Part 1, pretesting with two subjects, a series of stimuli given in Table 1 were cross adapted in the following manner. Subjects first tasted a test solution A after which they thoroughly rinsed their mouths with water. This provided the subjects with a preadaptation estimate of the intensity of the test solution. Immediately after rinsing with water the subjects held an adapting solution B in their mouths, swirling it around to ensure complete adaptation. After the sweet taste disappeared, which tended to be between 30–60 s, the subjects ejected the adapting stimulus from their mouths without a water rinse and quickly retasted test solution A, again estimating the intensity of its sweetness. The subjects noted whether the post-adaptation intensity was greater than, equal to, or less than, the pre-adaptation intensity.

The results of Part 1 are shown in Table 3. A positive sign indicates that the post-adaptation sweetness intensity of the test solution was perceived as less than its pre-adaptation sweetness. A negative sign indicates that the test solution tasted sweeter after adaptation, and an equal sign indicates that the test solution was considered to be unaffected by adaptation. A question mark indicates that the change was either too small for the subjects to

TABLE 3
DIFFERENCE IN PERCEIVED SWEETNESS AFTER ADAPTATION (RESULTS OF EXPERIMENT 2, PART 1: PRETESTING)

Test solution	Adapting solution										
	Galactose	Glucose	Fructose	Maltose	Sorbitol	Xylitol	Xylose	Acetosulfam	Ca cyclamate	Na saccharin	Rebaudioside
Galactose		++	+	++	++	++		--	==	?	++
Glucose	+		+	?	+	++		--	==	==	?
Fructose	++	++		++	++	++	++	--	==	==	++
Maltose	++	+	++		++	++	+	--	==	==	==
Sorbitol	++	+	++	++		++	+	--	?	?	--
Xylitol	++	++	++	++	++			--	==	?	?
Xylose	++	+	++	+				?	++	--	--
Acetosulfam	++	++	++	++	++	++	++		==	++	++
Ca cyclamate	++		+	++	+	++	++	==		==	==
Na saccharin	+	++	++	?	?	++	+	++	==		==
Rebaudioside		++	++	+	+	++	?	--	--	==	

For definition of +, −, = and ?, see text.

respond with confidence or the two subjects disagreed on the results. Overall, it can be seen that sugars were more effective as adapting stimuli than artificial sweeteners. The concentrations used for the stimuli were the same as those used in Experiment 1.

In Parts 2 and 3, each stimulus given in Table 1 served as an adapting solution while the rest of the stimuli were used as test solutions in the same manner as that described for Part 1. However, all combinations were tasted by at least 10 but not more than 45 subjects.

In Part 2, four different stimulus combinations were presented to each group of subjects. Twelve cups were arranged in a 3 by 4 matrix so that any three cups of a triad making up a column were part of a single stimulus combination arranged in an ABA design. Subjects were instructed to take the entire solution of the first cup A into their mouths and quickly rate its sweetness on a $5\frac{1}{2}$ in line labelled 'sweet' at one end and 'not sweet' at the other. After ejecting it immediately from their mouths and rinsing three times with water, the subjects then waited 4 min before holding the adapting solution B from the second cup in their mouths for 60 s. At the end of 60 s, they ejected the adapting solution from their mouths and promptly tasted test solution A again, from the third cup, rating it once more on the $5\frac{1}{2}$ in line. A water rinse (3 times) and 4-min inter-triad interval then followed before testing the next column or triad of solutions.

In Part 3 of the experiment certain combinations in Part 2 were retested. Only two triads were tested in a session, and the number of rinses and inter-stimulus intervals were increased. In addition, an effort was also made to determine the contribution of the water taste to the sweetness intensity of the test solution after adaptation. Adaptation to certain test stimuli is known to induce particular taste qualities in water. For example, McBurney and Schick[17] and Bartoshuk[18] have been able to produce sweet, sour, salty, and bitter qualities in water, following adaptation. These studies indicate that water, following adaptation to a stimulus with a bitter component, acquires a sweet taste. In Part 3, ratings of sweetness of water after adaptation to each of the test stimuli were made as well.

The mean for sweetness intensity of the test solution for each combination of the 8 stimuli before and after adaptation was calculated and the differences in these means are presented in Table 4. Data were combined from Parts 2 and 3. A positive magnitude of change indicates that following adaptation, there was a reduction in the taste solution's sweetness. A negative value indicates an enhancement. The statistical significance levels as determined by t-tests are given below the differences of the means. Table 5 shows the intensity of the sweet taste of water following adaptation to each of the stimuli as determined in Part 3.

It can be seen in Tables 3 and 4 that cross adaptation was consistently found to occur when a sugar was the adapting solution. However, when synthetic sweeteners were employed, enhancement frequently occurred.

Were it not for the complicated nature of the sweet quality, enhancement would immediately implicate more than one receptor site for sweetness. However, because many sweeteners have a bitter component, the argument is not that straightforward. As mentioned previously, water after adaptation to a bitter component acquires a sweet taste. Thus, the taste of a test solution after adaptation is the sum of the taste produced by the second compound and the water taste induced by adaptation to the first

TABLE 4
DIFFERENCE IN PERCEIVED SWEETNESS AFTER ADAPTATION (COMBINED RESULTS OF PARTS 2 AND 3 OF EXPERIMENT 2)

Test solution	Adapting solution							
	Acetosulfam	Aspartame	Ca cyclamate	Glucose	Neohesperidin dihydrochalcone	Na saccharin	D-tryptophan	Xylitol
Acetosulfam	—	10.0	8.0	5.4	−18.6	43.8 .001[a]	30.6 .01	25.2 .02
Aspartame	4.0	—	−7.4 .05	13.8 .05	−15.4 .05	−13.0 .05	40.0 .001	23.8 .01
Ca cyclamate	−11.0 .05	.7	—	20.6 .05	14.6 .01	4.2	−7.8	6.2
Glusose	−12.0 .20	−9.3	9.1 .20	—	−6.6	−17.4 .001	6.0	15.2 .10
Neohesperidin dihydrochalcone	−11.5 .20	6.6	−5.7	.8	—	−21.6 .001	15.8 .10	1.7
Na saccharin	29.7 .001	2.2	−15.4 .20	16.4 .10	−14.3	—	10.2 .20	2.0
D-tryptophan	−13.6 .10	36.0 .001	−14.0 .10	27.8 .05	−5.8	−5.6	—	22.0 .05
Xylitol	−9.2 .10	−1.4	9.2 .20	10.5 .20	−19.4 .001	−3.7	9.8	—

[a] Statistical significance levels as determined by *t*-tests

TABLE 5
WATER TASTE INDUCED THROUGH ADAPTATION TO THE STIMULI EMPLOYED IN THE STUDY

Adapting stimulus	No. of subjects	Arithmetic mean water taste
Acetosulfam	20	7·4
Aspartame	8	7·0
Ca cyclamate	18	12·2
Glucose	27	3·0
Neohesperidin dihydrochalcone	20	9·0
Na saccharin	10	11·8
D-tryptophan	9	2·6
Xylitol	35	9·6

stimulus, minus any cross adaptation between the two stimuli. Using this reasoning, the water taste was subtracted from the values in Table 4 giving the results listed in Table 6. It can be seen here that even when the results of Parts 2 and 3 are corrected for the water taste, enhancement still occurs for numerous adapting solutions.

Thus, it is suggested here that the fairly large, highly statistically significant inconsistencies in any one column of Table 4 (e.g. sodium saccharin) along with the fact that when water tastes are subtracted enhancement still occurs, as shown in Table 6, are not compatible with the existence of a single receptor site for sweetness. In Table 4 it is shown that adaptation to sodium saccharin resulted in a reduction of the sweetness of acetosulfam by 43·8 units on a 100 point scale. This change was significant at the 0·001 level. On the other hand, adaptation to sodium saccharin was shown to enhance the sweetness of neohesperidin dihydrochalcone by 21·6 units, also at a level of significance of 0·001. It is unlikely that such widely disparate results to adaptation for one compound are explicable by a single receptor.

The data in Table 6 were also analysed by an MDS procedure[19] such that the greater the degree of cross adaptation, the closer the sweeteners were located to one another in the space (see Fig. 6). It can be seen here that the pair, acetosulfam and sodium saccharin, tend to group together as do the pair aspartame and D-tryptophan. Members of a pair are more likely to share overlapping receptor sites.

TABLE 6
COMBINED RESULTS OF PARTS 2 AND 3 OF EXPERIMENT 2 AFTER CORRECTING FOR THE WATER TASTE

Test solution	Adapting solution						
	Acetosulfam	Aspartame	Ca cyclamate	Glucose	Neohesperidin dihydrochalcone	Na saccharin	D-tryptophan	Xylitol
Acetosulfam		17·0	20·2	8·4	−9·6	55·6	33·2	34·8
Aspartame	11·4		4·8	16·8	−6·4	−1·2	42·6	33·4
Ca cyclamate	−3·6	7·7		23·6	23·6	16·0	−5·6	15·8
Glucose	−4·6	−2·3	21·3		2·4	−5·6	8·6	24·8
Neohesperidin dihydrochalcone	−4·1	13·6	6·5	3·8		−9·8	18·4	11·3
Na saccharin	37·1	9·2	−3·2	19·4	−5·3		12·8	11·6
D-tryptophan	−6·2	43·0	−1·8	30·8	3·2	6·2		31·6
Xylitol	−1·8	5·6	21·4	13·5	−10·4	8·1	12·4	

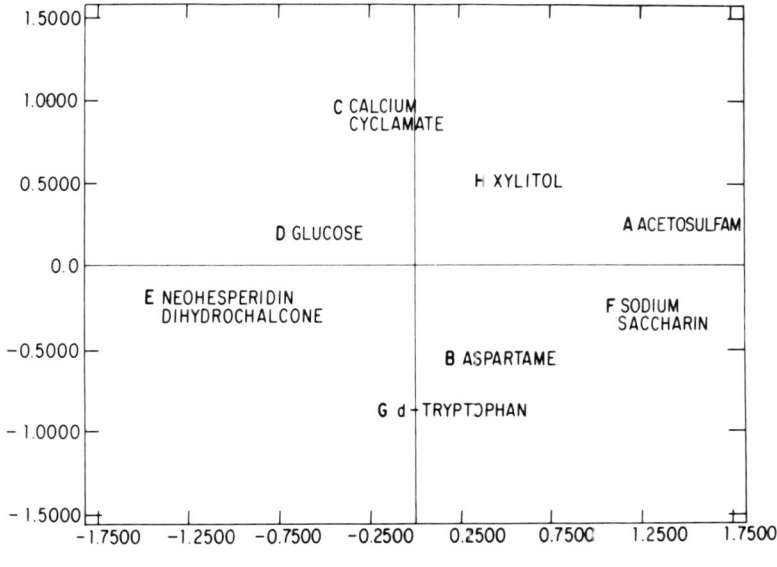

FIG. 6

Overall, the results in this paper illustrate two uses of multidimensional scaling: (1) to gain a geometric picture of the overall taste of sweeteners, and (2) to determine whether two stimuli are using the same sweet receptor site. One final point should be made here. In the space in Fig. 2, D-tryptophan and aspartame are located quite distant from one another but in Fig. 6 they are moderately close to one another. This is due to the fact that D-tryptophan has a strong bitter component. The three-dimensional space in Fig. 2 is based on overall quality. However, when sweetness alone was examined in the cross adaptation study, these two stimuli were found to share common receptor sites

In summary, multidimensional scaling can be helpful not only in understanding the phenomenological range of sweeteners but in delineating the underlying receptor properties as well.

REFERENCES

1. SCHIFFMAN, S. S., REILLY, D. A. and CLARK, T. B. (1979). *Physiol. & Behav.*, **23**, 1–9.
2. SCHIFFMAN, S. S. (1977). *J. Geron.*, **32**, 586–92.

3. SCHIFFMAN, S. S. and DACKIS, C. (1975). *Percept. Psychophys.*, **17**, 140–6.
4. SCHIFFMAN, S. S. and ERICKSON, R. P. (1971). *Physiol. Behav.*, **7**, 617–33.
5. SCHIFFMAN, S. S., MOROCH, K. and DUNBAR, J. (1975). *Chem. Senses Flav.*, **1**, 387–401.
6. SCHIFFMAN, S. S., MUSANTE, G. and CONGER, J. (1978). *Physiol. Behav.*, **21**, 417–22.
7. CARROLL, J. D. and CHANG, J. J. (1970). *Psychometrika*, **35**, 283–319.
8. TAKANE, Y., YOUNG, F. W. and DE LEEUW, J. (1977). *Psychometrika*, **42**, 7–67.
9. CARROLL, J. D. (1972). In: *Multidimensional Scaling: Theory and Applications in the Behaviour Sciences*, Eds Shepard, R. N., Romney, A. K. and Nerlove, S., Academic Press, New York.
10. BARTOSHUK, L. M. (1979). *Science*, **205**, 934–5.
11. FISCHER, R. (1967). In: *The Chemical Senses and Nutrition*, Eds Kare, M. R. and Maller, O. John Hopkins Press, Baltimore, 61–81.
12. GUTTMAN, L. (1968). *Psychometrika*, **33**, 469–506.
13. LINGOES, J. C. (1965). *J. Behav. Sci.*, **10**, 183–4.
14. MAZUR, R. H., SCHLATTER, J. M. and GOLDKAMP, A. H. (1969). *J. Am. Chem. Soc.*, **91**, 2684.
15. HOUGH, L. and PHADNIS, S. P. (1976). *Nature*, **263**, 800.
16. SCHIFFMAN, S. S., CAHN, H. and LINDLEY, M. (1981). *Pharma., Biochem., Behav.*, (in press).
17. MCBURNEY, D. H. and SHICK, T. R. (1971). *Percept. & Psychophys.*, **10**, 249–52.
18. BARTOSHUK, L. M. (1968). *Percept. & Psychophys.*, **3**, 69–72.
19. KRUSKAL, J. B., YOUNG, F. W. and SEERY, J. B. (1978). *How to Use Kyst-2A, a very Flexible Program to do Multidimensional Scaling and Unfolding*, Bell Laboratories.

Index

Absorption, 2, 102, 116
Accession efficiency, 279
Acetosulfam, 307
Acid conversion process, 85
Acid–enzyme conversion process, 85, 86
Acidity effects, 22–4
Additives, 4, 220
β-Adrenergic blockade, 235
AH–B system, 250, 280
AH–B–X system, 251
Aldopentose, 13
ALSCAL, 290, 292, 293
Amino acids, 30, 54, 55, 93
α-Amylase, 64, 76
β-Amylase, 64, 71, 76
Amylopectin, 78
Amylose, 77
Analytical techniques, 102
β-D-Arabinose, 13
Ascorbate oxidase, 26
Aspartame, 297, 307, 309
Aspergillus niger, 73, 121
Aspergillus oryzae, 72, 73
ATP, 159, 160, 164

Bacillus licheniformis, 70
Bacillus subtilis, 70, 72
Baking and bakery products, 91–2, 98, 119, 139
Beer, 99
Beverages, 137
Biochemical approach, 252

Biological activity studies, 266
Biophysical studies, 249–54
Biotechnology, 105–7
Biscuits, 139
Bitter/sweet interactions, 276–7
Bitterness, 276
Body weight response, 225–30
Boiled sweets, 42–3
Brown adipose tissue, 235–7
Brown sugars, 31–2
Bulking agents, 24–5

Cafeteria feeding in rats, 234
Cakes, 92, 139
Calcium phosphate, 220
Calories, 30
Canned food, 91
Canning, 98
Caramel, 31, 44–5, 211
Carbohydrates, 30, 39, 41, 46, 47, 50, 51, 56, 84, 92, 195–203, 226, 227, 237
  blood glucose elevation, effects on, 197
  chemical modification, 198
  dental caries, and, 206, 210, 213
  diet, and, 241–3
  long term studies, 200–2
  metabolism, 232–44
  physico-chemical modification, 197–9
  short term studies, 199–200
Cariogenicity. *See* Dental caries

Chemical approach, 249
Chemicals, 100
Chemistry of sweeteners, 7–15
Chirality, 10–12
Chlorosucroses, 297
Chocolate and chocolate products, 38, 39, 41–2, 138, 141
Cholesterol, 201
Citric acid, 175
Colour, 31
Computer control, 25
Concentration effects, 22–4
Confectionery, 37–48, 89–90, 98
  consumption, 38
  history, 37–8
  manufacture, 39–40
  miscellaneous types, 45–6
  nutritive sweeteners, functions of, 40–7
  see also under specific types
Corn syrups, 215
Coupling sugar, 218

Dairy products, 99
DE value, 85–9
Debranching enzymes, 78
Dental caries, 2, 29, 30, 167, 205–24
  prevention of, 219–21
  sugar consumption, and, 210
Dextrorotatory structures, 13
Dextrose, 41–3, 46, 93–8
  anhydrous, 97
  applications, 98–100
  caloric value of, 95
  chemical reactivity, 94–5
  crystalline, 93, 94
  digestion and utilisation, 95
  fondants, 44
  manufacture, 96–8
  properties of, 93–4
Dextrose equivalent (DE), 85
Diabetes, 176, 177, 232, 237, 244
Diabetic food, 116
Diet, 30, 31, 117, 172, 175, 176, 180, 181, 186

Dietetic foods, 92, 142
Digestion, 181–9
Disaccharides, 225
Disease states, 3, 196
Drugs, 100

Economics, 18
EEC directives, 38, 39, 106–7
Electrophysiological approach, 255, 262
Embden–Meyerhof–Parnas glycolytic pathway, 157, 159
Energy
  balance, 20
  conservation, 20
  sources, 19–20, 116
  storage, 34–6
  value, 102
Enzyme
  supplements, 61
  systems, 50, 67, 186
Equilibrium relative humidity, 47
Escherichia coli, 123

Faeces analysis, 201
Fermentation, 92
Flavour, 31
Fluoride concentration, 29
Fondants, 43–4, 46
Food
  consumption, 177
  preference, 174–6
Frozen desserts, 99
β-D-Fructofuranose, 9, 10, 13, 14
β-D-Fructopyranose, 8
Fructose, 9, 41, 46, 100, 133–44, 196, 214, 228, 276
  applications, 137–42
  characterisation, 134–5
  food systems, in, 134
  relative sweetness, 136
  sweetness, 135–7
Fruit juices, 100
Fudges, 45, 46
Fuzzy patterns, 254–60

Galactose, 117, 120, 214
α-L-Galactose, 13
β-Galactosidase, 126
Gastric secretion, 181
Gelatinisation temperature, 28
β-Glucanase, 56, 59, 64
β-D-Glucopyranose, 12
Glucose, 117, 120, 199–201, 228
Glucose isomerase enzyme, 101
Glucose-6-phosphate, 95
Glucose syrups, 40, 41, 43, 45, 61, 85–9, 196, 198, 200
 applications, 89–93
 definition, 85
 fermentability, 88
 general characteristics and properties, 86–9
 manufacture, 85
 molecular properties, 88
 pH values, 87
 reducing characteristics, 89
 solids content, 86
 sweetness, 87
 textural characteristics, 88
 viscosity, 88
Glucuronate–xylulose Touster's cycle, 157
Glycolytic cycle, 95
Grapefruit juice, 138

Health drinks, 92
Hexitols, 146
High fructose corn syrup, 100
High fructose glucose syrups, 100–3
 analysis, 102
 applications, 103–6
 comparative sweetness, 102–3
 manufacture, 101
 physiological properties, 102
 second generation, 104–5
 storage and use, 103
Horecker–Racker pentose phosphate pathway, 157
Hormones, 228, 241
Humidity, 32–3
Hydrogen bonding, 250

Hydrogenated sugars, 145–70
 metabolism, 156
 physico-chemical properties, 153–4
 production, 152
 relative sweetness, 154
 technological applications, 169
 utilisation in food industry, 168
Hydrogenation, 198–9
Hydrolysis products, 214–15
Hygroscopicity, 41, 86, 87

Ice cream, 92–3, 104, 140
Immunoreactive insulin (IRI), 187
INDSCAL, 290, 292, 293
Infant foods, 92
Insulin, 174, 187, 188, 201, 232, 237
Intensity/time sweetness analysis, 277–9
Interaction energy, 269–70
Invert sugar, 46

Jams, 90, 104, 140
Jellies, 90, 140

Klebsiella aerogenes, 78
Kluyveromyces lactis, 120
Krebs' cycle, 157, 159

Lactitol, 124–6, 218
 characterisation, 124–5
 potential applications, 125–6
 preparation, 125
Lactobacillus avei, 126
Lactobacillus bifidus, 124
Lactobacillus fermenti, 126
Lactose, 41, 46, 110, 214
 anhydrous, 112
 characterisation, 111–17, 123
 crystalline, 112
 hydrolysis, 120–1
 manufacture of, 118

Lactose—contd.
  nutritional considerations, 114–17
  physiological considerations, 114–17
  relative sweetness, 114
  solubility behaviour, 113
  sorption properties, 113
  transformations, 111
  utilisation of, 118
Lactose hydrolysates, 120–2
  properties of, 121–2
  uses of, 122
Lactulose, 122–4
  applications, 123–4
  manufacture, 123
  physiological effects, 123–4
Laevorotatory structures, 13
Legislation, 38, 39, 100, 220
Lemonade, 137
Lycasin, 152, 169, 201, 202, 217, 218
Lycasin syrups, 154–6

Magma, 96
Maillard reactions, 45, 119
Maillard-type reactions, 40
Malt extract, 53, 54, 56
  alternative production, 60–1
  modern production, 61–4
  wort analyses, 67
Malt replacement, 61
Malt syrups, 49–67
  composition, 52–5
  high extract, 68
  laboratory and plant scale production, 64, 65
  process improvement, 66–7
  traditional production, 55
Malting biochemistry, 55–8
Malting problems, 58
Maltitol, 153, 167, 199–202
Maltose, 41, 46
Maltose syrups, 50, 67–79
  high conversion, 73
  improved production, 70–2
  low-glucose-containing, 70

Maltose syrups—contd.
  production using single enzyme process, 77
  traditional production, 69–70
Mannitol, 153
D-Mannitol, 149–50, 162–3
Mannose, 7
β-D-Mannose, 14
Massecuite, 96
Meat curing, 26
Meat products, 99
Melting point, 154
Metabolic rate, 228–9
Metabolism, 181–9
Milk, 110, 117, 120, 214
Moisture level, 32–4
Monosaccharides, 93, 95, 225, 228
Multidimensional scaling, 287–310
Multiple enzyme process, 85, 86

Neurophysiological studies, 254–60
Nitrogen balance, 186
Nutrients, 172
Nutrition, 1–3, 29, 171–93

Obesity, 176, 232, 236–7
Off-flavours, 276
Oligosaccharides, 7, 71, 75
Oral factors, 173–7
Oral stimulation, 181–9
Oxidation, 198

Packing, 98
Palatability, 18, 32, 176, 177, 179, 186
Palatinit, 218
Pancreatic endocrine secretions, 187
Pancreatic exocrine secretions, 182, 185, 187
Pancreatic polypeptide secretions, 185
Pentitols, 146
pH effects, 73, 85, 87, 101, 117, 206, 209, 212–15, 219

Phenylthiourea, 294
Physiological effects, 172, 173,
    195–203
Pickles, 99, 104
Polyhydric alcohols, 297
Polysaccharides, 30
Portal systemic encephalopathy
    (PSE), 124
Postingestional factors, 174
Postingestional stimulation, 185
PREFMAP, 293, 297
Prepared dry mixes, 98
Preservation, 25–6
Preserves, 104
Pronase E, 261
Protamine zinc insulin, 238
Psychological studies, 254–60
Psychophysical approach, 257
Puddings, 140
Pullulanase, 78

Q sugar, 24

Reducing sugars, 89

Saccharides, 297
Sauces, 104
SDS acrylamide gels, 252
Sensory Measuring Unit for
    Recording Flux ('SMURF'),
    277
Sensory mechanism, 172
Shell-graining, 89
Sodium saccharin, 307
Soft drinks, 91, 100, 103, 138
Sorbitol, 27, 46, 147, 150–3,
    157–62, 169, 201, 202, 219
Spoilage, 34
SSAI technique, 294
Starch, 83–108, 214–15
    history, 84
Stereochemistry, 8
Streptococcus mitis, 126
Streptococcus mutans, 29, 30, 211,
    218

Streptococcus salivarius, 126
Streptococcus sanguis, 126
Structural studies, 264
Structure–activity relationship,
    249–54
Structure–taste relationships, 8–15
Sucrose, 17–35, 43, 46, 196, 210,
    219, 227
    bulk function, 24–5
    caramelisation of, 31
    consumption, 18–19
    cost, 27
    preservative, as, 25–6
    purification, 32
    purity, 26
    structure, 28
Sugar
    beet, 20
    cane, 20–1
    consumption, 1, 18, 210
    elimination, 219
    prices, 147
    production, 147
    substitution, 215–18, 220
Sugar/water interaction, 279
Surfactants, 280
Sweet taste, 8, 176, 177
Sweet taste receptor mechanisms,
    247–73
    biophysical studies, 249–54
    individual differences, 260–70
    neurophysiological studies,
        254–60
    psychological studies, 254–60
    structure–activity relationship,
        249–54
Sweeteners, 1, 5, 21, 215–18
    chemistry of, 7–15
    interindividual sensitivity
        differences, 260–70
    non-sugar, 215–18
Sweetness, 4–5, 18, 21–4, 40, 87,
    102–3, 135–7
    evaluation, 287–310
    hydrogenated sugars, of, 154
    intensity, 284, 303
    intensity/time analysis, 277–9
    measurement, 22, 24

Sweetness—contd.
  nutritional significance. *See*
    Nutrition
  response to, 173
  standard, 23
Synergism, 137, 275–85
  intensity/time analysis, and,
    277–9
  localised concentration, and,
    279–80
  sugar/water interaction, and, 279
  surfactants, and, 280–4

Taste cells, 4
Temperature effects, 22–4, 34
Thaumatin, 280
Thermogenesis, 232–4
  biological implications, 243–4
  diet-induced, 233–43
  non-shivering, 235–41
  rats, in, 234
Thirst quenchers, 100
Toffees, 44–5
Tooth decay, 29
Toxicity tests, 5
Transformation process, 24

Trichlorogalactosucrose
  (1',4,6'-trichlor-1'4,6'-
    trideoxygalactosucrose),
    218
Triglycerides, 201, 232
D-Tryptophan, 307, 309

Urine analysis, 201

Ventromedial hypothalamus, 241
Viscosity, 25, 40, 281
Viscosity effects, 88
Vitamins, 30, 174

Water solubility, 154
Wine, 99
Wort carbohydrates, 52
Wort/malt extract, 53

Xylitol, 146, 148, 153, 164–6, 169,
  216, 219, 276, 297
Xylose, 297

Yogurt, 141